SD226 Biological psychology: exploring the brain
Science: Level 2

Book 3
Exploring the Brain

This publication forms part of an Open University course SD226 *Biological psychology: exploring the brain*. The complete list of texts which make up this course can be found on the back cover. Details of this and other Open University courses can be obtained from the Student Registration and Enquiry Service, The Open University, PO Box 197, Milton Keynes MK7 6BJ, United Kingdom: tel. +44 (0)845 300 60 90, email general-enquiries@open.ac.uk

Alternatively, you may visit the Open University website at http://www.open.ac.uk where you can learn more about the wide range of courses and packs offered at all levels by The Open University.

To purchase a selection of Open University course materials visit http://www.ouw.co.uk, or contact Open University Worldwide, Walton Hall, Milton Keynes MK7 6AA, United Kingdom for a brochure. tel. +44 (0)1908 858793; fax +44 (0)1908 858787; email ouw-customer-services@open.ac.uk

The Open University
Walton Hall, Milton Keynes
MK7 6AA

First published 2004. Second edition 2006. Reprinted 2007

Edited, designed and typeset by The Open University.

Printed and bound in the United Kingdom by Halstan Printing Group, Amersham.

ISBN 978 0 7492 1432 6

2.2

The paper used in this publication contains pulp sourced from forests independently certified to the Forest Stewardship Council (FSC) principles and criteria. Chain of custody certification allows the pulp from these forests to be tracked to the end use (see www.fsc-uk.org).

SD226 COURSE TEAM

Course Team Chair

Miranda Dyson

Academic Editor

Heather McLannahan

Course Managers

Alastair Ewing

Tracy Finnegan

Course Team Assistant

Yvonne Royals

Authors

Saroj Datta

Ian Lyon

Bundy Mackintosh

Heather McLannahan

Kerry Murphy

Peter Naish

Daniel Nettle

Ignacio Romero

Frederick Toates

Terry Whatson

Multimedia

Sue Dugher

Spencer Harben

Will Rawes

Brian Richardson

Other Contributors

Duncan Banks

Mike Stewart

Consultant

Jose Julio Rodriguez Arellano

Course Assessor

Philip Winn (University of St Andrews)

Editors

Gerry Bearman

Rebecca Graham

Gillian Riley

Pamela Wardell

Graphic Design

Steve Best

Sarah Hofton

Pam Owen

Picture Researchers

Lydia K. Eaton

Deana Plummer

Indexer

Jane Henley

Contents

BASIC CELL BIOLOGY

1.1 Introduction

A chapter on cell biology in a biological psychology course may seem out of place, but it is necessary for two reasons. The first reason, mentioned at the beginning of Book 1 (Book 1, Section 1.2.2), is that one way of describing the brain is in material terms, i.e. in terms of its chemical constituents. You have already met some of the key chemical constituents of cells (e.g. proteins and chromosomes) and their functions. This chapter explores in a little more detail the relationship between these constituents. The second reason for a chapter on cell biology, also introduced in Book 1 (Book 1, Section 1.3.2 and Section 2.4.3), is that the key component of the nervous system, the neuron, is a cell. Knowing how cells function can lead to possible causal explanations of brain mechanisms that might otherwise be missed.

The cell, the basic unit of life, out of which all organisms are constructed, was described in Book 1 (Section 2.4.1). A typical human cell is virtually indistinguishable from a single mouse, frog, fly, or worm cell. Each cell is engaged in a number of processes, including respiration (the production of ATP, the energy currency of the cell) and metabolism (the synthesis of new molecules and the destruction of old molecules). These vital processes require proteins, which are considered in Section 1.3. Section 1.4 deals with the structure of DNA (the genetic material) and Section 1.5 deals with the role of DNA in protein production.

Proteins and DNA are macromolecules, and the next section, Section 1.2, briefly describes these and the two other main types of macromolecules that are together the principal chemical constituents of cells.

This chapter necessarily simplifies cell biology, but the information presented here is sufficient for the needs of this course.

1.2 The macromolecules

There are four types of **macromolecules**: carbohydrates, lipids, proteins and nucleic acids. Each macromolecule is composed of a number of smaller, component molecules; proteins, for example, are composed of amino acids. The component molecules are all obtained from the diet. Usually, whatever food is eaten, it is in the form of macromolecules, which have to be broken down (i.e. digested) into their chemical components. These are then transported in the bloodstream to cells, where they are metabolized, i.e. converted into other chemical components or reassembled into the macromolecules and structures of the host.

A few chemicals, e.g. water and salt, are obtained directly from the diet, i.e. they enter the body directly without the need to be broken down and reassembled.

Carbohydrates and lipids require only a brief word here. Carbohydrates, such as starch, can be broken down to their chemical components, called monosaccharides, one of which is the crucial energy yielding molecule glucose, that fuels respiration. Cells that have high energy demands, such as those in the brain, the nervous system and muscles need to be supplied with lots of glucose. Lipids are present in large quantities in cell membranes and are also the main constituent of myelin, the electrically insulating sheath which surrounds nerve axons. The chemical components of lipids, fatty acids, are the building blocks for steroid hormones (e.g. testosterone). Both carbohydrates, stored as glycogen, and lipids, stored as fat, can be utilized at a later time as an energy source.

Proteins need to be considered in more detail than do carbohydrates or lipids. Proteins are important for a number of reasons, for example, certain proteins, called enzymes, control metabolism. Enzymes dictate which chemical conversions can occur, and hence which chemical constituents can be made. A simple example will illustrate the point. Most animals have the enzyme necessary to convert glucose into vitamin C. However, we do not have the appropriate enzymes in our cells, so we cannot make vitamin C – a failing we share with other primates and with guinea pigs. (Vitamin C is essential for life, so we must obtain it directly from the diet.) Enzymes and other proteins are considered further in Section 1.3.

Nucleic acids also need to be considered in more detail. One nucleic acid, deoxyribonucleic acid, **DNA**, is found in the cell nucleus, and comprises the genetic material. DNA is the chemical component of the long-term information stores more usually referred to as chromosomes and genes. The other nucleic acid, ribonucleic acid, **RNA**, serves a number of functions, one of which is as a short-term information store, relaying information from the nucleus to the cytosol (see Figure 1.1a). This important function determines which proteins are to be assembled in the cytosol. DNA contains the full complement of an organism's genes and, it is the genes that determine which proteins an organism has. We do not have the proteins, the enzymes, needed to make vitamin C, because we do not have the necessary genes. The structure and function of DNA is considered in Sections 1.4 and 1.5.

1.3 Proteins and enzymes

Proteins have many and varied functions, all of which are necessary for life. Some proteins are contractile. Muscle, for instance, consists principally of the proteins actin and myosin. Some proteins contribute to the shape of individual cells and structures (e.g. blood vessels, and microfilaments). Others act as messengers in the organism, being released from cells in one part of the body and affecting the activity of cells in another part. Insulin, for example, is released from the pancreas and assists the movement of glucose from the bloodstream into the cells. Some proteins are receptors which detect these messengers. They are situated in the plasma membranes of cells and thereby influence cells.

Enzymes, as stated above, are one particular sort of protein. They control the rates at which the myriad chemical conversions in cells take place; most chemical conversions in cells would be very slow, or would not occur at all without enzymes. One important neurotransmitter in the brain, dopamine, illustrates the role of enzymes very well. Dopamine itself is not a protein. However, it is made from tyrosine, which is an **amino acid** and amino acids are the component molecules of proteins. The conversion of tyrosine into dopamine requires two different enzymes, each controlling a separate step in the conversion process. Without those enzymes, dopamine could not be produced.

Every species of living organism has its own characteristic set of proteins, called the *proteome*. Humans, for example, can make somewhere between 50 000 and 70 000 different proteins in total. Some of these proteins are very similar or even identical to those of other organisms, including flies and worms. But only humans have the human proteome. That does not mean that all the cells of your body produce all these proteins all the time; far from it. You produce a different sub-set of the proteome at different times during your life. Furthermore, different tissues produce different sub-sets of the proteome. It does appear that proteins are made to order; so that the right protein is in the right tissues at the right time. The way in which protein production is controlled is a key element of change in all its forms, i.e. growth, development, healing, learning and memory. Section 1.5 describes protein production.

Proteins are made of amino acids joined one to another into strings of varying lengths. (Short strings, of four or five amino acids, are sometimes called peptides, whilst longer strings are called polypeptides. Both are proteins.) Twenty or so different amino acids are commonly found in living organisms and, by arranging different numbers of them in different orders, so different proteins can be made. (It is worth noting that a string of only six amino acids could come in one of 64 million forms! The number of possible proteins is enormous.) The exact order in which the amino acids are joined together is dictated by the genetic material by means of the genetic code – the topic of Section 1.5.

However, a protein is not simply a string of amino acids: a protein has a unique three-dimensional shape. The shape arises because a particular string of amino acids folds up on itself in a particular way. The precise shape of a protein is often decisive in the functioning of that protein: if the protein is the wrong shape, its functioning will be impaired, or worse, it will not function at all. A mistake in joining the amino acids together into a string, so that just one is of the wrong kind or in the wrong place or missing, can alter the shape of the folded protein and result in a protein which simply cannot do its job. It is therefore, literally, vital that proteins are made accurately.

Cells producing a particular sub-set of the proteome are said to be specialized. For example, a cell may be specialized to produce and release insulin into the bloodstream. To do so it must contain the molecular machinery to synthesize insulin, to secrete insulin and to control both of these processes in response to signals from elsewhere in the body. Other cells may be specialized to produce the fat-rich myelin around nerves in vertebrates, whilst others may be the nerves themselves. However, humans do not produce cells specialized in the production of silk; nor do spiders have specialized cells to produce myelin. The reason for these differences lies in the DNA of humans and spiders respectively. It is now time to consider this remarkable molecule.

1.4   The structure of DNA

The genetic material comprises several lengths of DNA (Figure 1.1a), each length being formed from two strands of DNA joined together (Figure 1.1b).

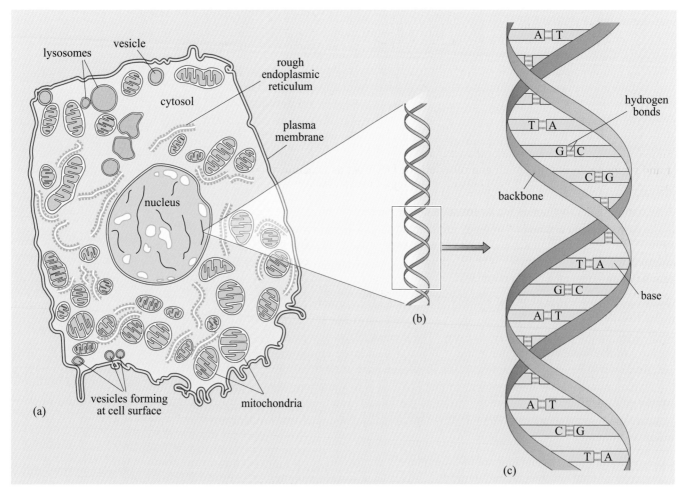

Figure 1.1   Schematic drawing of a typical animal cell and DNA showing different levels of detail. (a) Schematic drawing of a typical animal cell showing the genetic material, DNA, constrained within the nucleus of the cell. (b) DNA showing two DNA strands. (c) Illustration of part of a molecule of DNA. A = adenine, T = thymine, C = cytosine and G = guanine.

Each strand of DNA consists of a 'backbone', with component molecules called nucleotide bases attached to it. There are four such bases; adenine, thymine, cytosine and guanine. The two strands are held together by bonds between these bases, always adenine to thymine and cytosine to guanine (Figure 1.1c). The order of the bases in the genetic material of any one species is remarkably stable though there is some variation between individuals. The exact sequence of bases in one of the strands (the template strand, described below), determines two things. First, it determines the exact sequence of bases in the other strand (A to T, C to G remember). Second, the exact sequence of bases in the template strand determines the exact sequence of amino acids in the proteins it produces, and, as you have seen, the exact sequence of amino acids in a protein determines whether or not that protein can function, by affecting its shape and composition. It is the precise sequence of bases in the template strands of DNA that is the genetic material and that comprises the information that is passed from generation to generation in the egg and sperm cells.

Virtually all the cells of an organism contain a complete and exact copy of the genetic material found in the organism at conception, the zygote. A pristine copy of the genetic material in each cell allows the cell to continually break down its current stock of proteins and replace them with new proteins produced from the DNA template. This turnover of proteins is necessary to ensure the current set of proteins is appropriate and functioning. DNA is not subject to such a level of 'turning over' and generally remains unchanged for the life of the cell.

Not all the DNA codes for protein. This is certainly true of the other, i.e. non-template strand, but is also true of the template strand. Some DNA of the template strand seems to do nothing at all and other bits of DNA act as attachment points for the proteins that control, and the enzymes that undertake, the decoding process.

A section of DNA that codes for a protein is called a gene. A gene may be a single continuous section of DNA, but it may not be, a complication that we return to later. All the genes contained within the cells of an organism are called the **genome** or genotype of that individual. As there are many thousands of proteins, and each protein is coded for by a particular gene, it follows that there must be many thousands of genes. These genes are arranged so that there are very many genes in one template strand of DNA. The genes are strung together in a precise order and are interspersed with non-coding sections of DNA, sometimes disparagingly termed junk DNA. (Around 1.5% of human DNA codes for proteins.) Each different template strand comprises a set of different genes.

1.5   The function of DNA

The amino acids that form proteins are joined together with the help of ribosomes on the rough endoplasmic reticulum in the cell (Figure 1.1a). The mechanism by which the information encoded in the DNA (in the nucleus) is brought together with the ribosomes involves an intermediary molecule, mRNA. The intermediary molecule is ribonucleic acid (RNA) and it carries the **m**essage from the template strand in the nucleus to the ribosomes, hence the name messenger RNA (**mRNA**).

Unlike DNA, mRNA molecules are not present in the nucleus all the time but are made specifically to order using a short section of DNA as a template. The process of using the template to make a complementary RNA strand begins with a part of the DNA double helix (Figure 1.2a) unwinding and separating its two strands (Figure 1.2b), in response to molecular signals. RNA is made along *one* of these strands, the template strand, by following a very similar pairing procedure to that used between the two strands of the DNA molecule (Figure 1.2c). In DNA, adenine is always paired with thymine, and cytosine is always paired with guanine. In RNA, the pairing system is the same except that thymine is replaced with uracil. So to each thymine molecule exposed on the template strand of DNA, an adenine is attached; to each cytosine, a guanine is attached; to each guanine, a cytosine is attached and to each adenine, a uracil is attached (Figure 1.2c).

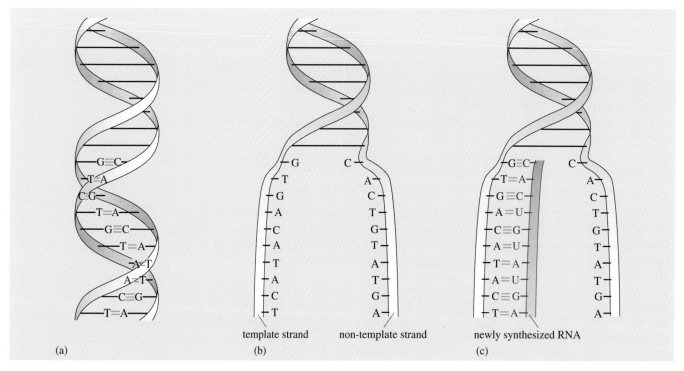

(a)   (b)   (c)

template strand   non-template strand   newly synthesized RNA

Figure 1.2   The transcription of a DNA base sequence into an RNA base sequence. (a) Part of a DNA double helix showing a few base pairs. (b) The same DNA double helix but unwound to reveal two sets of bases. (c) The unwound DNA molecule showing the synthesis of a molecule of RNA on the template strand.

A protein, RNA polymerase, is responsible for matching DNA bases to RNA bases and joining the RNA bases together. RNA polymerase attaches to the start sequence of the exposed DNA and, using nucleotide bases present in the nucleus, builds a chain of bases *complementary* to the DNA bases on the template strand; this process of synthesizing RNA on a DNA template is called **transcription**. Note that the non-template strand of DNA has the same sequence of bases as the RNA product, albeit with uracil replacing thymine. When the RNA polymerase reaches a particular sequence of three bases in the exposed DNA, known as the stop sequence, transcription is complete, the single stranded RNA molecule detaches from the DNA template strand, allowing the DNA to rewind itself. The molecule of RNA produced is the *primary transcription product*, not mRNA.

As was mentioned earlier, only some sections of the template strand of DNA code for protein, i.e. they are expressed or 'turned on'. These sections are called **exons**. Other sections of the template strand do not code for proteins, these sections are called **introns**. When the template strand is transcribed, both introns and exons are copied, so that the RNA molecule, the primary transcription product, contains copies of the wanted exons and the unwanted introns. Whilst the RNA molecule is still in the nucleus, the unwanted introns are cut out by special proteins, restriction enzymes, and the remaining sections of wanted RNA are joined back together again by other proteins, splicing enzymes. The result of this cutting and splicing is functional mRNA (Figure 1.3).

The cutting and splicing of the primary transcription product occurs after transcription and so is called post-transcriptional modification.

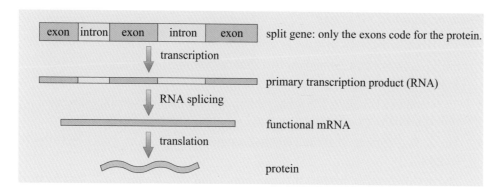

After post-transcriptional modification, the newly synthesized mRNA molecule is transported out of the nucleus with the aid of specialized and specific transporter proteins. mRNA becomes attached to the ribosomes and the process of **translation** begins, where the ribosomes match the sequence of bases in the RNA to the appropriate amino acids in the cytosol, join those amino acids together and produce a protein.

It was mentioned earlier that a section of DNA that codes for a protein is a gene. So there is a gene for the protein mentioned in Book 1, Section 1.3.3, arginine vasopressin. However the word 'gene' is actually a collective term, like the terms cake and rose; there are a number of different types of cake and a number of varieties of rose. In the case of the gene for arginine vasopressin, five different varieties of the gene are known in humans, each with a slightly different DNA sequence. Each variety of a gene is an **allele**, so there are five alleles for human arginine vasopressin. Everybody carries the gene for arginine vasopressin in their cells, but different people may carry different alleles for it.

At some point after the mRNA has been translated, and there may have been several ribosomes working on it at the same time so many copies of the protein could have been made, the mRNA is released from the ribosome and is broken down to its component bases ready to be recycled. The synthesis and degradation of mRNA means that there is a continuous *turnover* of mRNA in response to the demands of the cell and the signals it is receiving.

1.6 Summary of Chapter 1

The cell is the basic unit of life, it is surrounded by a membrane that protects the contents from the external environment. The cell carries out many processes, including metabolism and respiration, which are controlled by special proteins called enzymes. A key metabolic process is the turnover of proteins. The assembly of proteins requires enzymes, ribosomes and mRNA, made from a DNA template. The genetic material DNA is contained within the nucleus. A section of one DNA strand is copied by a process called transcription to yield a primary transcription product. The primary transcription product undergoes modification to remove introns. After this post-transcriptional modification, the resultant mRNA is transported out of the nucleus and into the cytosol. In the cytosol, mRNA becomes attached to ribosomes where its sequence of bases is translated into a sequence of amino acids to produce the gene product, a protein. After translation, the mRNA is released from the ribosome and is broken down into its constituent bases, adenine, cytosine, uracil and guanine.

Transcription in 12 steps

1 The DNA in the nucleus is the genetic material.

2 Each length of DNA comprises two strands, a template and a non-template strand. (Illustrated in Figure 1.4a.)

3 Each template strand of DNA has coding and non-coding sections. The non-coding sections are sometimes unfairly referred to as junk DNA.

4 The coding sections (i.e. the genome) comprise start lengths, genes and stop lengths.

5 The genes comprise lengths that may be wanted, called exons, and lengths that are not wanted, introns.

6 The gene is transcribed as a length of RNA, the primary transcription product. (Illustrated in Figure 1.4b.)

7 All introns are removed, and some exons may also be removed from the primary transcription product during post-transcriptional modification. (Illustrated in Figure 1.4c.)

8 The resulting RNA, messenger RNA, can then be transported out of the nucleus.

9 Once outside the nucleus the mRNA attaches to ribosomes and is translated into protein.

10 The mRNA is in due course broken down into its component bases.

11 Most genes are subject to variation.

12 Any variant of a gene is called an allele. Alleles are variations on a gene. (Illustrated in Figure 1.4d.)

non-template

ATGATGATGATGTHEQUICKCLEWBROWNFOXJUMPSOVERGARDENGNOMETHELAZYDOGCCGCCGCCGCCG

ZGTZGTZGTZGTGSVJFRXPXOVDYILDMULCQFNKHLEVITZIWVMTMLNVGSVOZABWLTXXTXXTXXTXXT

template

(a) Step 2   The above two lines of letters represent DNA, comprising a template strand (i.e. the lower one) and a non-template strand (i.e. the top one). (Note: the full alphabet has been used to reveal patterns; DNA is more realistically represented as a four-letter code. In this version of the code A pairs with Z, B pairs with Y, etc.)

THEQUICKCLEWBROWNFOXJUMPSOVERGARDENGNOMETHELAZYDOG

(b) Step 6   The line of letters above represents the primary transcription product of the gene embedded in the strand of template DNA. (Yes, it does look similar to the non-template strand.) Exons are now shown in blue, introns are now shown in red.

THEQUICKBROWNFOXJUMPSOVERTHELAZYDOG

THEQUICKFOXJUMPSOVERTHELAZYDOG

THEFOXJUMPSTHELAZYDOG

THEBROWNFOXJUMPSOVERTHEDOG

(c) Step 7   Post-transcriptional modification could result in a number of different mRNAs, and hence proteins, depending on how the restriction enzymes present in the nucleus cut up the primary transcription product.

THEQUITECLEWBROWNFOXJUMPSOVERGARDENGNOMETHELAZYDOG

THEQUITECLEWSLOWOXJUMPSOVERGARDENGNOMETHELAZYDOG

THEQUICKCLEWBROWNFOXJUMPSOVERGARDENGNOMETHEDOG

THEQUICKCLEWBROWNFOXLAMPSOVERGARDENGNOMETHELAZYDOG

(d) Step 12   This gene has undergone several changes, i.e. mutations, over time to produce five alleles, shown here as primary transcription products. (The fifth is shown in Figure 1.4b.)
Any one individual will have two of the five alleles in each of their cells, and these two alleles could be different or identical.

Figure 1.4   Illustrations of stages 2, 6, 7 and 12, using the full alphabet, not just CGAT, to make the patterns clearer.

Learning outcomes for Chapter 1

After studying this chapter you should be able to:

1.1 Recognize definitions and applications of each of the terms printed in **bold** in the text.

1.2 Explain the importance of proteins.

1.3 Explain the relationship between proteins and DNA.

1.4 Outline the key stages in transcription.

Questions for Chapter 1

Question 1.14 *(Learning outcome 1.1)*

The following statements are misleading; they are nearly correct, but are actually wrong. Change, add or remove *one* word from each sentence to make them correct.

Note: the important part is to see why the statements are wrong; there may be several changes that can correct the sentences, but only one is given in the answer.

(a) Everyone has exactly the same gene for arginine vasopressin.

(b) The proteome is the full complement of proteins in a cell.

(c) The primary transcription product is mRNA.

Question 1.2 *(Learning outcome 1.2)*

Give four examples of functions that proteins perform.

Question 1.3 *(Learning outcome 1.3)*

In Section 1.4 the phrase 'the exact sequence of bases in the template strand' was used.

(a) What is the template strand?

(b) What does the exact sequence of bases determine?

(c) Why is the exact sequence important?

Question 1.4 *(Learning outcome 1.4)*

The following sentence is correct, but very general. Replace the words in italics to make a more specific statement about translation.

Translation is the process by which *organelles* match the sequence of *component molecules* in a *macromolecule* to the appropriate *component molecules*, join those *component molecules* together and produce a *macromolecule*.

STUDYING THE BRAIN: TECHNIQUES AND TECHNOLOGY

2.1 Introduction

The aim of biological psychology is to be able to understand the brain mechanisms of behaviour, and to that end it is necessary to study the brain. This chapter sets out some general principles about studying behaviour, and some specific techniques used to study the brain. Some of the techniques you have already met, and many of the techniques will be revisited in later parts of the course. This chapter serves to bring them together and put them in the general context of studying the brain. The chapter begins with the beginning of the scientific process.

2.2 The scientific process

Science is not immune to the political, social, financial and personal factors that influence all human endeavour. However, there is a core scientific process which represents good scientific practice, and it is this core scientific process that is outlined in this section.

The scientific process cycles through the stages of observation, hypothesis construction, experimentation, data generation, data analysis, data interpretation and hypothesis reconstruction. The skills associated with the scientific process, as they apply to this course, are taught as part of the Investigative Strand. In this section we consider some of the principles associated with observation, objectivity and experiments, beginning with observation.

2.2.1 Observation

Science begins with an observer noticing some character or quality that, to the observer, is of interest and awakens their curiosity. An initial observation, for example the sudden, short duration, episodes of sleep in a friend (this example is considered further in Section 2.5.3), may set in train a whole series of further observations and studies. The subject matter of this course is the brain, and you have seen from previous chapters some of the things that the brain can do and some of the behaviours that are studied. Indeed, more than likely you have noticed things yourself about individuals and puzzled over their abilities: what makes a chess champion, a train spotter, an octogenarian marathon runner? Where the observation is of just a single individual, it is called an anecdote, or case report, depending on who does the observing and where they make their observations known. Your story about your mother catching a precious vase when most of the time she couldn't move quickly, or the same story appearing in a magazine or newspaper, is, however true, an anecdote. A similar story about a patient with Parkinson's disease catching a precious vase when most of the time they couldn't move, told by a trained scientist (e.g. a psychologist or neurologist), written about in medical journals or reported to medical conferences is a **case report**. The distinction between the case report and the anecdote rests partly on the experience of the observer, partly on the authority (i.e. professional versus lay person) of the observer and partly on the *objectivity* of the observer; objectivity is considered in the next section. Case reports have their place in the investigation of human behaviour; two of the areas of the brain involved in language were identified because of an individual studied in detail by Paul Broca and another individual studied in detail by Karl Wernicke. Similarly, the unfortunate Phineas Gage and H.M., both of whom lost areas of their brain, informed the debate about which areas of the brain were involved in emotional restraint and memory respectively. (The contribution of each of these individuals was considered in Book 1, Chapter 1.) A. R. Luria was intrigued by the remarkable memory of a journalist he knew and wrote a detailed account of his abilities in *The Mind of a Mnemonist* (Luria, 1986). Oliver Sacks has produced similar, but less extensive case reports (Sacks, 1985). In 2001, Tim Lawrence, a stuntman afflicted with Parkinson's disease, persuaded a BBC TV crew to film a case report. A stuntman is dependent on good motor control and Tim Lawrence had lost this. However, the TV crew filmed him performing various gymnastic routines, including a trampolining session. He claimed that it was the drug MDMA (methylenedioxymethamphetamine) that had temporarily restored his motor control.

Each case report is remarkable in its own way and sheds light on the way the brain works. But each is also unique. It is therefore not always appropriate to extrapolate from observations on these individuals to the workings of the brains of other individuals.

◆   What are the two main problems of extrapolating from observations on one brain-damaged individual (e.g. Phineas Gage) to how the brain works in other individuals?

◆   One of the main problems is that observations of brain-damaged individuals tell you how their brain copes in the *absence* of the damaged area, which is not direct information about what that area does in the undamaged brain. The other main problem is that it is difficult to establish precisely the extent of the damage.

◆   What is the assumption that is made in saying that patients such as H.M. and Phineas Gage provide valuable insight into the workings of the human brain?

◆   The assumption that must be made is that other individuals must work in a similar way; that their brains must be wired in a similar way to the individual in the case report.

This assumption is taken to be valid by most people who study the brain. Bear in mind though that it is an assumption, so don't be surprised when you read later in the course (e.g. Section 3.8.2) how the brains of people differ, both in their structure and in their workings.

Case reports may be informative, they are invariably interesting and they may provide the basis of a theory, but they do not provide good evidence from which to generalize about the population. Rather more than one observation is needed to make general statements and to substantiate a theory – case reports need corroboration. That corroboration can take the form of other case reports, or of results of another kind of study supporting the inference of the case. For example, the role of Broca's area in language, first established with brain-damaged patients, has been corroborated with brain scanning techniques, considered later (Book 5, Chapter 2).

Case reports do feature in this course, but the bulk of the subject matter is the product of careful, systematic observation of many individuals, which means that the great body of information reported in this course meets two important criteria. The first is that the observations fit within accepted theories and the second is that the observations have been duplicated or reported on several occasions. There may well be occasions when one or other, or even both, of these criteria is not met, in which case the observation is said to be pushing the limits of science or 'at the cutting edge', but for the most part the course deals with 'accepted' knowledge. An important example from history will illustrate the distinction between cutting edge science and accepted knowledge.

The anastomosis paradigm

The principal ideas in this debate were introduced in Book 1, Box 2.1. Scientists in the 1800s, with their sophisticated new microscopes, really wanted to know what the vertebrate brain was made of. They meticulously prepared the brain tissue, making the slices as thin as possible. They then carefully stained the material with the new selective stains of the day and then painstakingly drew what they saw when they viewed the samples through a microscope (e.g. Figure 2.1).

Of particular interest were the neurons. Each neuron was generally thought to be physically joined (by neurofibrils) at its extremities to the extremities of other nerve cells (Figure 2.2). The hypothesized physical link formed an uninterrupted tube, an anastomosis, between nerve cells.

It was believed that the function of the physical link was to allow one nerve cell to communicate with another. The presumed physical connection fitted in with the prevailing theory that nerve cells communicated with each other using electricity. Then, towards the end of the 1800s, along came Ramón y Cajal. He could not see the joins between nerve cells in his preparations of brain tissue, so he drew the neurons as separate entities (Figure 2.3).

The argument raged: the body of evidence provided by other scientists seemed to refute Cajal. Furthermore, Cajal could provide no explanation of how nerve cells communicated if they were separate from each other; electricity does not smoothly cross gaps. Any textbook of the time would have presented nerve cells as joined and would have presented Cajal's observations as, what we now call, cutting edge science (or as an anomaly or as a mistake). Anastomosis was the accepted knowledge of the time. Accepted knowledge, like the joined nerve cells, is not necessarily right, but it is knowledge that holds sway in the scientific community at any given time. Much of what you will read about in this course is accepted knowledge. However, because it could be wrong, it is important that you know what makes it 'accepted' and what would be required to overturn it. Hence this chapter on the scientific process and the book on experimental design (Book 2).

Figure 2.1 Microscope from the end of the 19th century.

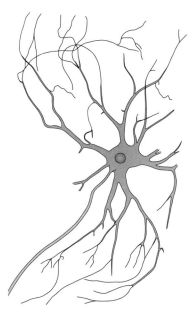

Figure 2.2 Drawing of a neuron by Deiters in 1865.

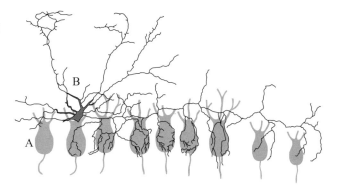

Figure 2.3 Drawing of a basket neuron (B) by Cajal connecting with nine other neurons (e.g. A).

2.2.2 Objectivity

The rules of scientific observation today require that observers are **objective**, by which it is meant that the observer records only what is actually observed or measured; that the observer is not influenced by their own beliefs, expectations or emotional involvement. Imagine for example, evaluating the reading performance of two children for a prize, one of whom is your offspring; objectivity under these circumstances would be compromised by your emotional involvement with one of the contestants and the consequences of your evaluation, i.e. tears of joy of your child. Without any criteria against which to evaluate the readers, your evaluation would be *subjective*. Using specific criteria, e.g. in the current example, diction, and the appropriateness of pauses, would move you from being subjective towards being *objective*. (Subjectivity and objectivity can be regarded as being at opposite ends of a single scale.) But you would remain the parent of one of the contestants and hence could always be accused of emotional involvement and lack of objectivity. Objectivity, even under the best of circumstances is very difficult to achieve, because of the simple fact that the observer is involved in the observation. Observers often have a vested interest in what they are observing and a knowledge of the underlying theory, and they may be swayed accordingly. Objectivity, then, is an ideal to which all observers strive, but which few achieve. Objectivity can be achieved by using mechanical and electronic measuring devices, which is why they are sought. It can also be more closely approximated if the observer is counting or measuring 'something', and knows nothing about why they are doing so (in the sense that they are disinterested and have no 'attachment to the results'). Also, the better the 'something' is defined, the more accurate and objective the observer will be.

Were the microscopists of the 1800s being objective? We have to give them the benefit of the doubt and assume that they were being as objective as they could. They were, after all, operating at the very limits of their technology, and presumably drew what they believed they saw. What is remarkable is that Cajal was willing to risk ridicule by seeing something different to everybody else, or rather, by failing to see what everybody else purportedly saw.

Another feature revealed by this story of early microscopists is that science operates within both a theoretical and a technical context. (There is also a political context about what science is funded and what science is permitted, but that is beyond the scope of the present discussion.) The theoretical context required electricity to flow through the nervous system without being interrupted, so there could not be gaps in the network. At the time, there was no known mechanism that would allow information to cross gaps: the scientific 'story' was better without Cajal's gaps! Theoretical considerations still exert their influence today, though, of course, it is not known which theories are influencing which observations, and it will not be known until hindsight is brought to bear. It was suggested in the early 1950s that the Earth's magnetic field was a stimulus to which organisms could respond. This idea was rejected by the scientific community, at least in part, because of the lack of a plausible mechanism by which organisms could detect magnetism. Although behavioural evidence that birds could detect the Earth's magnetic field began to accumulate in the 1960s, it was not clear cut and a mechanism remained elusive. The controversy continued until 1975, when R. P. Blakemore provided conclusive evidence of the mechanism; he reported that bacteria which clumped together near a magnet contained magnetite (Blackmore, 1975) – a substance subsequently sought and found in other organisms, including birds.

Or consider the following, written in 2001 by Sian Lewis, the editor of the influential *Trends in Neurosciences*:

> Over the past few years, the classic idea that no new nerve cells are born in the adult mammalian brain has finally and conclusively been refuted by the scientific community. Yet, the first indications that neurogenesis [the birth of new neurons] occurs in the brain of adult mammals were obtained using light and electron microscopy over two decades ago. Why this went unrecognised is described in a personal account by the researcher who pioneered those studies: Michael Kaplan.

The technical context of Cajal and his contemporaries, was that the microscopes of the day were not really good enough to see the tiny gaps between nerve cells. Indeed, microscopes that rely on visible light (light microscopes) are still not good enough. Incontrovertible evidence of the gaps had to wait until the advent of the electron microscope in the 1950s (Figure 2.4; and also in Figure 2.24 in Book 1).

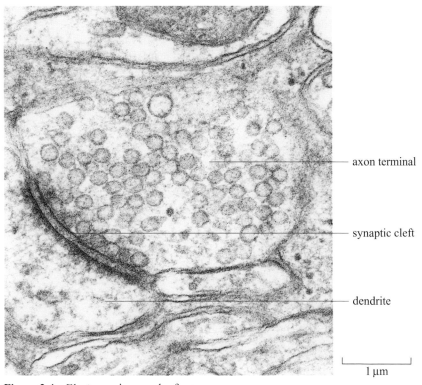

axon terminal

synaptic cleft

dendrite

1 μm

Figure 2.4　Electron micrograph of a synapse.

It remains the case that technology is limiting. Scientists are always wanting to push technology; today's technology merely stimulates scientists to ask questions that need tomorrow's technology to answer. For example, in the 1970s and 80s, scientists used radioactive material and dead brains (yesterday's technology) to reveal that certain areas of the brain might be involved in different tasks. The process was slow and inaccurate, but now today's technology (scanning) allows scientists to measure accurately which small areas of the brain are active during which behavioural tasks, and to do so in real time – that is, as it happens. (There is more on these topics later in the chapter.)

2.2.3 Research methods

There are many research methods that inform biological psychology. They may involve:

- questionnaires, where people tick boxes to agree or disagree with printed statements, or items presented on a computer screen;

- interviews, where people are questioned about their feelings or motives;

- associations between naturally occurring events;

- observations under different, but naturally occurring conditions;

- controlled intervention.

These different research methods are extremely valuable and widely used, and they have whole courses written about them. No one particular research method has precedence and the material presented in this course has been obtained with the use of many different methods. All these research methods are used in investigations and are often called experiments. This is the *broad* definition of experiments. However, it is only possible in this course to consider one research method in detail, and that is the one in which there is controlled intervention. This is the *narrow* definition of an experiment. Experiments that fit the narrow definition can be distinguished from the other research methods by the presence of controlled intervention, which you will recall as the independent variable. Intervention can be characterized as 'let's see what happens if …', but the reality is that the intervention is carefully considered in the light of contemporary theories and very precisely controlled. Experiments also demand replication. Replication requires that exactly the same intervention is performed on several different occasions, and that is only possible if the intervention is strictly controlled. Cajal was able to provide supporting evidence for his view of neurons by intervening with the material. He reasoned that if neurons were joined one to another, then destroying one neuron would have an effect on others. He was able to destroy individual neurons and observed no effect on adjacent neurons. His conclusion was that neurons were each separate from one another.

◆ Are case reports experiments, within the narrow definition given above?

◆ Case reports are not experiments with an independent variable because there is no controlled intervention and there is no replication.

These sorts of experiments feature in Book 2, where the details of good experimentation are set out.

2.3 Asking questions

When interviewed in the media about their work, scientists often give very complex answers to simple questions. This is not because they are by nature evasive or incoherent people, but because, in science, simple questions often require complex answers. This was amply illustrated in Book 1, Chapter 1 where a variety of questions and explanations in biological psychology was explored. In this section we pick up on one particular explanatory principle for further consideration.

2.3.1   Cause and effect

Causal explanations involve making clear statements about cause and effect. Once such statements have been made, then predicting the consequences that small changes in the cause have on the effect can be made, which ultimately allows for accurate, corrective intervention. Science makes quite good predictions when things have a neat, sequential, causal relationship. But the subject matter of this course is not like that; neat, sequential causal relationships are very much the exception. There are two reasons for the difficulty in establishing causal relationships in biological psychology. The first reason is that there are a number of different timescales in operation. For example, one theory about the cause of the mental condition known as schizophrenia is the neurodevelopmental hypothesis. Schizophrenia is rarely diagnosed before adulthood, the symptoms generally becoming apparent in the late teens or twenties. Yet the neurodevelopmental hypothesis suggests that abnormalities are already present in childhood, caused by events occurring very early in development. (The neurodevelopmental hypothesis is considered in detail in Book 6, Chapter 3.) Proving such a causal relationship is extremely difficult because of the huge time interval between the formation of the brain in the womb and the symptoms of schizophrenia appearing. That large time interval allows any number of other intervening factors to exert their effects and be the cause, or contribute to it. Proving causal relationships when events are much closer together in time (e.g. between a blow to the head and immediate memory loss) is much easier, though not without its problems. Causal relationships are sought in biological psychology and become more robust as the timescale reduces from years to weeks to days to hours, but the degree of interaction between factors and the extensive network of checks and balances (feedback loops) underpinning homeostasis, makes establishing those relationships very difficult.

The second reason for the difficulty in establishing causal relationships in biological psychology is that there are a number of different levels to be considered (Book 1, Section 1.4). For example, a blow to the head may cause memory loss, but it can also cause physical damage to the brain and intense electrical activity. So rather than just regarding the blow to the head as painful and mildly traumatic, it is also necessary to ponder the physical damage to cells in the brain and the unusual electrical activity of neurons. These additional considerations extend the causal chain of events from 'the head' to cells and their activity. The original neat, causal relationship becomes obscured by the intervening levels.

These intervening levels also lend themselves to observation and experimentation, though the techniques and technology required are entirely different from those required to observe behaviour.

Summary of Section 2.2 and 2.3

There is a core scientific process which if followed precisely would yield accurate, irrefutable data. Unfortunately, science takes place in a social context which means that a variety of social factors affect the scientific process, and this is particularly true in biological psychology. Observations are affected by observers, so great efforts must be taken to render the observation objective. Even with objective observations, establishing reliable cause and effect relationships remains problematic, not least because there are different timescales and different levels to accommodate in any explanation.

2.4   Techniques and technology: unpacking the brain

There are a limited number of ways of studying the brain. In essence, all that is required is a method of stimulating the brain and a method of recording the consequences of that stimulation. This section begins by considering the normal situation in which the brain receives stimulation from the sense organs and the response is a behaviour of some sort. However, it is possible to bypass the sense organs and stimulate the brain directly, with electrical or chemical stimuli, and also to pre-empt the behavioural response by recording electrical and chemical activity in the brain. These latter two scenarios are considered after the normal situation. (Delayed behavioural responses, i.e. those occurring months or years after stimulation, are considered in Chapter 3.) Only the main techniques are described to give you an indication of the possibilities, so this account is not comprehensive; nor is it very detailed. Greater detail of specific techniques will be provided where required in the course.

2.4.1   The normal, living brain

Perhaps the most widely used method of studying the brain is to present natural stimuli of one sort or another, i.e. sound, light, smell, taste and touch, and record the resultant behaviour. This is sometimes called the 'black box' approach because the brain is not examined directly. The stimuli are detected by the normal sensory systems of the organism, and the way the brain uses that sensory information is *deduced* from the response, from the behaviour. This approach does not address the question of *how* the brain actually does its job, but it does explore what the brain is able to do.

Measuring and observing people's response to particular stimuli is predominantly the realm of psychology. The range of potential tasks is enormous, ranging from physically interacting with stimuli to thinking about them or describing feelings as a consequence of those stimuli. The effect of subliminal images (different faces) on the pleasantness of a drink was mentioned in Book 1, Section 1.1.6, but memory, skill acquisition, multitasking and childhood experiences (i.e. developmental influences) are all fair game in evaluating what the brain is able to do.

The stimuli are those to which we can respond, whether consciously or not, and they may be presented in such a way as to assist the task set or hinder it. The task can be anything from remembering a list of words to deciding whether one taste is preferred over another. Finally, the response may be categorized into one of four main types:

- a quick motor response, e.g. pressing a lever;
- a physiological response, e.g. sweating or pupil dilation;
- a verbal response to questions about the task, e.g. a questionnaire or introspection;
- a replay of the original stimuli, e.g. recall of lists, redrawing a picture.

Whatever the task, the basic premise is the same: that it is possible to deduce something about the functioning of the brain from knowing the stimulus or stimuli that preceded a particular response. A small sample of tasks that can be performed over the Internet is given in Table 2.1.

Table 2.1 Examples of experiments available on the Web.

Laboratory experiments	Research experiments
covert attention study	emotional Stroop experiment
facial recognition experiment	figurative language
learning and memory	face recognition
line motion experiment	infant communication
maze experiment	moral choice
mirror drawing experiment	perception of gender with inverted faces
perception of gender	personality and response speed
political poll experiment	reaction times between different athletic sports
reaction time visual	Stroop and recognition study
lateralized Stroop experiment	two-day facial recognition study
word recognition experiment	word–face associations

One well used example, the **Stroop** effect, involves words and colours. Each word is the name of a colour (e.g. red) and it is presented in coloured ink or on a colour monitor. The ink colour can be compatible with the word (e.g. the word 'red' printed in red ink) or the ink colour can be incompatible with the word (e.g. the word 'red' printed in blue ink). The investigator sets the participant the task of, for example, naming the colour of the word, whilst ignoring the meaning of the word. Generally, incompatible words take longer to read than compatible words. The duration of the task gives an indication of how the brain processes information.

The black box technique is also used with animal subjects. One well known piece of technology that exploits the black box technique is the Skinner box, which you met in Book 1, Section 1.4.2.

As well as presenting stimuli to an organism's sense organs, such as the eye in the Stroop example, it is also possible to stimulate the organism directly using electricity or chemicals, e.g. drugs such as alcohol. These direct methods of stimulation have an effect on the nervous system, but their effect does not require sense organs and is not restricted to sensory pathways. In this way, direct stimulation can be said to bypass the sensory systems.

Chemicals, e.g. drugs, can be ingested (i.e. eaten or drunk), or injected, either into the bloodstream or directly into the brain. Drugs in the bloodstream are transported throughout the body and brain, so their place of action is not controlled. Even if the only part of the body which can detect and respond to injected chemicals is the brain, as is the case with mood altering and stimulant drugs such as valium and amphetamine, the site(s) of action within the brain remain uncontrolled.

◆ What prevents many chemicals in the bloodstream from coming into contact with neurons in the brain?

◆ The tight junctions between endothelial cells that form the blood–brain barrier. (See Book 1, Section 3.5.2.)

An alternative and very precise method of delivering a chemical to the brain is with **microinjections**. Very thin stainless steel needles can be pushed through an opening in the skull, into the brain and used to inject tiny quantities of chemicals into a ventricle, or a very small area of the brain. If required, the needles can be attached to minipumps to ensure continual infusion. It is even possible to inject chemicals to influence individual neurons, using very fine glass tubes, called micropipettes. This technique is called **microiontophoresis**.

Electricity can also be used to stimulate the brain directly. The small electrical contacts that are used to deliver the electricity are called *electrodes*, whether they are on the skull, on the surface of the brain or inside the brain. If the electrodes are on the surface of the skull, then any electrical impulse is both reduced in strength and dispersed by the skull before it can affect the brain. Such external electrodes therefore can deliver only a rather imprecise impulse. Also, the skull hinders the passage of electricity through it, a property known as impedance. To overcome the impedance of the skull, the electrical impulse needs to be relatively large. A large, imprecise electrical impulse is not a suitable stimulus for studying the finer workings of the brain. However, it is a suitable stimulus if the intention is to create excessive neuronal activity, a seizure. A violent electrical storm in the brain may seem an odd thing to want to induce; it is after all what happens in epilepsy, and epilepsy is very unpleasant. The reason for inducing the seizure, usually just in the non-speech dominant hemisphere, is to treat depression. This treatment is called electroconvulsive therapy, ECT.

Electrodes on the surface of the brain can deliver a small, localized impulse, but only to areas of the brain exposed by the removal of the overlying skull. This technique of **direct electrical cortical stimulation (DECS)**, to give it its full title, has been used to map the function of some parts of the surface of the brain and has proved particularly valuable where electrical stimulation produces specific movements of parts of the body, e.g. the fingers. More precision is possible, and smaller electric currents can be used, when the electrodes are very thin tungsten wires. These **microelectrodes** can be pushed deep into the brain through a suitable opening in the skull. The tungsten wire is insulated except for its extreme tip where a small electric current can stimulate a small, localized area of the brain or even individual cells.

Microelectrodes are increasingly being used in the treatment of movement disorders, such as the rigidity and tremor caused by Parkinson's disease. The treatment is called **deep brain stimulation (DBS)** and has proved an effective alternative to drug therapy. DBS uses an implanted microelectrode to deliver continuous high-frequency electrical stimulation to either the thalamus, or the globus pallidus, one of the structures comprising the basal ganglia (Book 1, Figure 3.11). Permanently implanted microelectrodes are also used to stimulate the spinal cord. Low-frequency electrical stimulation of the dorsal column (Book 1, Figure 3.21) is used to treat severe and intractable pain.

One difficulty of using micropipettes and microelectrodes is in knowing where the tip is within the brain. This difficulty can be partly overcome by positioning the head in a stereotaxic frame. Markings on the frame enable the tip to be guided to a designated three-dimensional location in the brain. Alternatively, the tip can be watched and visually guided to its location using imaging machines, e.g. X-ray scanners. A third option is to mark the position of the tip, for example by releasing a small amount of dye from the micropipette. This technique has the considerable limitation of requiring the direct examination of the brain at autopsy, where the dye can be seen and the position of the tip established.

The positioning of the electrodes in DBS is done empirically; the patient is awake and the electrode tested, i.e. an appropriate amount of current is passed through it, until the location that the patient says best reduces the symptoms (tremor or rigidity in the case of Parkinson's disease) is found. It's a bit like getting someone to scratch an itch on your back, but you can't give them directions on where to scratch (the patient does not know where the electrode is), only how successful the scratching is (how much the symptoms are reduced).

Electricity can be induced in wires by using magnets. The same principle can be applied to the nervous system, where neurons, or more specifically, their axons, are the 'wires'. By applying a suitable, focused pulse of magnetism it is possible to induce electrical activity in a small group of neurons in the brain. This procedure of **transcranial magnetic stimulation (TMS)** is non-invasive and is completely reversible.

Just as the normal sensory input to the brain can be bypassed with electrical or chemical stimulation, so the normal behavioural consequences of stimulation can be pre-empted, by using electrical and chemical activity. Chemicals and electricity can be used to stimulate the brain, and they can also be used to measure the activity of the brain. Blood, urine and cerebrospinal fluid can all be sampled and examined to see if they contain specific chemicals. The chemicals are of two types. Some chemicals are products secreted by the brain to influence other parts of the body, e.g. hormones and neurotransmitters. Other chemicals, known as breakdown products, or metabolites, give an indication of which chemicals have been used in the brain, just as wrappers in the bin give an indication of what someone has been eating. The breakdown products provide two pieces of information. They reveal which chemicals have been used and also, by their quantity, the extent to which they have been used.

To record chemical activity within the brain requires two very fine tubes or cannulae (singular **cannula**), one within the other, to be inserted through a hole in the skull. The inner cannula carries a salt solution into the brain and the outer cannula carries the salt solution out of the brain. At the tip of the two cannulae is some special tubing, called *dialysis* tubing, which allows chemicals in the fluid surrounding it to enter the cannulae. This happens because the salt solution entering the brain contains none of the chemicals of interest found in the fluid surrounding the cells of the brain. Those chemicals in the brain fluid surrounding the dialysis tubing are all moving and some just happen to move, to diffuse, across the dialysis tubing into the salt solution. The salt solution that leaves the brain also contains whatever chemicals have been picked up from within the brain. The process of collecting chemicals from within the brain using dialysis tubing is called **microdialysis**. Chemical analysis can then be applied to the outcoming fluid to determine which chemicals are present around the dialysis tubing.

The brain can be stimulated by electricity, but the brain also generates electricity in quantities that vary with its activity. Electrical activity can be recorded using electrodes. A pair of electrodes, attached close together on the scalp with an electrically conducting gel, can be used to detect the cumulative effect of the tiny voltage changes (a few microvolts) generated by nearby neurons. The signal is amplified and fed into a computer. It is easier to visualize what the computer does by describing the apparatus the computer has replaced. So, rather than the signal being amplified and fed to a computer, consider that the signal is amplified and fed to a device, a galvanometer, which converts the signal into the sideways movement of a pen. A strip of paper moves at a constant speed under the pen. As the current measured by the electrodes changes, so the pen moves back and forth across the paper, creating a long wiggly line, a brain wave. Usually there are several pairs of electrodes on the scalp, each feeding a different pen, or channel. There can be up to 40 channels on one multichannel recorder. The result, from the pen and from the computer, is the **electroencephalogram**, or **EEG** (Figure 2.5).

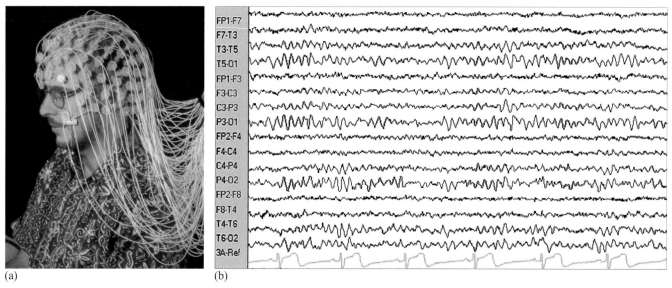

Figure 2.5 EEG array and record. (a) The electrodes attached to the skin send their signals to the recording apparatus (not shown) via a net of fine wires. (b) Each line on the EEG record has as its source two recording electrodes, indicated on the left (e.g. T4–T6; T6–O2, etc.)

The electrical signals can be accurately timed. If a stimulus, e.g. a sound, is applied at a known time in relation to the EEG recording, then the effect of the stimulus on the EEG can be deduced. This process underlies the **event-related potential (ERP)**. In fact the deduction involves a lot of computation, and repeated application of the stimulus. Those electrical events which usually accompany the stimulus are enhanced, whilst those which only occasionally accompany the stimulus are reduced. Eventually, an averaged response emerges from the wealth of electrical signals, and a large, unmistakeable, slow wave, the event-related potential, becomes clear. The wave is identified by its time in milliseconds after the stimulus, say 200 ms, and whether it is in the negative (N) i.e. downwards, or positive (P) i.e. upwards direction (Figure 2.6). The P200 is a fairly standard (i.e. regularly observed) ERP shown on Figure 2.6 as P2 (the second positive peak). ERPs have been used to investigate many types of cognitive processes, including memory, language and attention, face recognition in children, as well as degenerative disorders, such as Alzheimer's disease.

◆ What particular recording problems does the EEG have to overcome?

◆ The problems associated with the skull. In exactly the same way as the skull disperses and attenuates an electrical impulse used to stimulate the brain, so the skull attenuates and disperses the weak electrical signals generated by the brain.

◆ How might the electrical interference of the skull be minimized?

◆ The electrical interference of the skull can be minimized by using microelectrodes.

Microelectrodes can be used to record the electrical activity of relatively small groups of cells, or even of individual neurons, inside the brain. Micropipettes (sometimes called glass microelectrodes) have such fine tips that they can also be inserted into individual brain cells and used to record their activity.

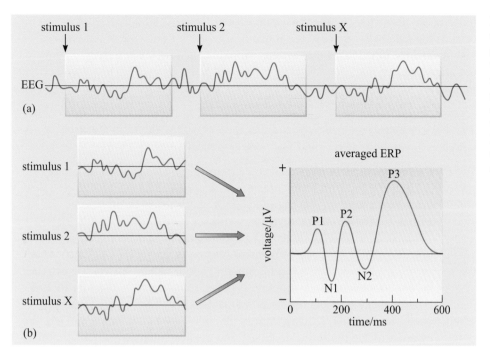

Figure 2.6 A schematic illustration of the ERP. (a) The stimulus is repeated a large number of times and the EEG recorded. The initial records from a single pair of electrodes show no discernible pattern. (b) Computer analysis for consistent EEG responses transforms the initial records into the ERP (see the text for a description). (P1, P2, P3 refer to the first, second and third positive peaks.)

◆ Three methods of recording electrical activity have just been mentioned. What were they?

◆ The EEG, recording the activity of small groups of cells using a microelectrode and recording the activity of single neurons, using a glass microelectrode.

These three methods of recording electrical activity can be used in real time and whilst the subject is active and moving around. They can also detect changes occurring in a very short space of time, milliseconds, which is close to the timescale in which neurons operate.

There is an additional way to measure the electrical activity of the brain, but the device that is able to measure it requires the subject to keep very still. Any electrical current necessarily generates a magnetic field, and this is true of nerve cells even though the electric current and the magnetic field they generate is tiny. However, by placing an array of supercooled, highly sensitive detectors around the skull, it is not only possible to detect but also to identify the source of the magnetic fields generated by activity in the brain: this is the principle of the **magnetoencephalogram**, or **MEG**. The MEG allows the localized areas of electrical activity that arise in the brain in response to particular stimuli to be identified.

There are other devices that measure brain activity and that require the subject to be very still, but these devices do not measure electrical activity: they measure blood flow. The brain requires a continuous supply of blood to provide both energy, in the form of glucose, and oxygen. The brain can neither store these substances nor use alternatives. About 20% of the heart's output of blood goes to the brain. The progress of the blood through the brain though is not a flood, but a carefully controlled irrigation. The network of blood vessels in the brain is under very sensitive, localized control, such that each area of the brain receives only the amount of blood that it requires: all areas of the brain do not receive equal amounts. Those areas that receive the most blood during a particular task (e.g. presentation of an auditory stimulus, thought or manual manipulation) and hence are in receipt of the most glucose and oxygen, are deemed to be the most active. (This activity, which takes place within cells, is the metabolic activity mentioned in Chapter 1.) The size of the area of activity and the intensity of that activity though should be taken only as a guide to the importance of that area for whatever task is being assessed. The relative importance of the crowd and the players at a sports meeting, compared to their size and energy consumption, is a useful analogy here.

Two devices measure blood flow and distribution within the brain; positron emission tomography (PET) and magnetic resonance imaging (MRI).

The **PET scan** – Radioactive material emits small, high-energy particles called positrons. Each emitted positron interacts with a nearby electron, resulting in the annihilation of them both and the production of gamma-rays (a form of electromagnetic radiation like X-rays, but of higher energy). The gamma-rays disperse in equal and opposite directions from the point of positron–electron annihilation and can be detected by suitable sensors. A computer can reconstruct the source of the gamma-rays using information about which sensors are activated and when. The process of reconstruction is called tomography – the T in PET. A patient or participant has a small dose of radioactive material injected into their bloodstream. The material is transported in the blood around the body and into the brain. The areas of the brain that command the greater volumes of blood produce the most gamma-rays and it is these areas that are computed and displayed by the PET scan (Figure 2.7a).

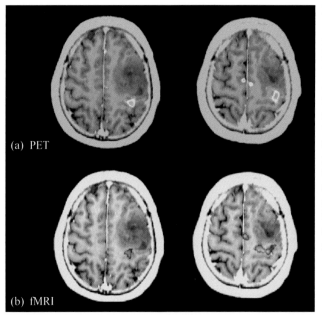

(a) PET

(b) fMRI

Figure 2.7   Four brain images taken whilst a patient finger tapped with the left hand. The front of the brain is towards the top of the figure. Activity is apparent in the right hemisphere. (a) PET scan at two horizontal levels within the brain and (b) fMRI scan at two levels of the same patient. Colour has been added to show areas of high activation. Note that in fMRI and PET the activity of the brain is recorded before and during tapping (see the text opposite).

The **MRI scan** – Any charged particle that spins has magnetic resonance. Protons have a positive charge and they spin all the time. These properties mean that all atoms and molecules have magnetic resonance because they contain protons. The technique depends not only on the fact that magnetic resonance can be detected, but also on the fact that different molecules have different magnetic resonance. Two molecular components of the blood are particularly interesting in this regard; they are variants on haemoglobin, the molecule that makes blood red. Haemoglobin with oxygen attached is called oxyhaemoglobin, whilst haemoglobin that has no oxygen attached to it is called deoxyhaemoglobin. The magnetic resonance of deoxyhaemoglobin is different from that of oxyhaemoglobin. When blood is diverted to particular areas of the brain, the ratio of oxy- to deoxyhaemoglobin will change, and this change can be detected by the sensors. Computer wizardry (see below) produces an image (the I in MRI) of where the change in ratio occurs.

One advantage of MRI over PET is that it does not require patients or participants to be exposed to radioactive substances. The absence of any potentially dangerous exposure means that, unlike the situation with PET, images of the same person performing the same task can be repeatedly produced. However, MRI scanners are noisy and the patient is confined in a small space, having to use mirrors to see out of the scanner, and has to remain very still. The noise, constriction and motionlessness make for an uncomfortable experience. The computer wizardry referred to above, effectively subtracts the images produced when the participant is not performing the task from the images produced when the participant is performing the task. What remains is a composite image showing the difference in blood flow between when the participant was and was not performing the task. The difference in blood flow is usually represented as a colour scale on images and is assumed to indicate where the brain carries out the particular function; this is **functional magnetic resonance imaging, fMRI** (Figure 2.7 here, and Book 1, Figure 3.23).

◆　Does the PET scan also require 'before' performing the task and 'during' performance of the task images?

◆　Yes, the PET scan results from a comparison of 'before' and 'during' images, the former being subtracted from the latter.

Figures 2.7, 2.8 and 2.9 compare the brain area involved in finger tapping as localized by fMRI and PET, or finger flexion as localized by direct electrical cortical stimulation (DECS) and transcranial magnetic stimulation (TMS). It is clear that there is a reasonable amount of consistency between the different techniques in identifying the active brain area during finger tapping.

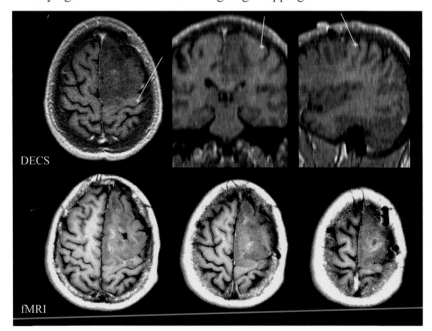

Figure 2.8　Comparison of fMRI generated whilst the patient was finger tapping and direct electrical cortical stimulation. The upper row illustrates the position of the stimulating electrode which elicited contraction of the small hand muscles used in finger tapping.

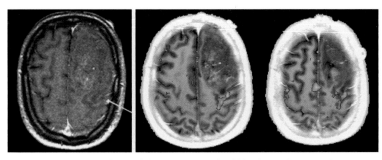

The MRI scan, tuned to detect the magnetic resonance of hydrogen, generates high-resolution, three-dimensional images of the brain. These images reveal the anatomy and structure of the brain in some detail (Figure 2.10). This is structural MRI.

Figure 2.9   Comparison of fMRI generated whilst the patient was finger tapping and transcranial magnetic stimulation. The pointer in the left-hand image indicates the position of the TMS which elicited contraction of the muscles used in finger tapping. The two images on the right are fMRI scans at two levels in the same patient.

Scans that use X-rays are generally just referred to as CT (computerized tomography) scans and, although common, the resultant images are two-dimensional, and of comparatively low resolution. The sole function of the MRI scan and the CT scan is to show structure, so they differ from other techniques mentioned in this section.

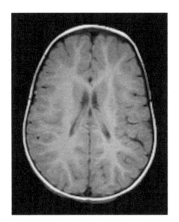

Figure 2.10   Structural MRI scan of a normal brain.

◆   What is it that the PET, MRI and CT scans cannot do that the other techniques mentioned in this section, i.e. electrodes, TMS, can do?

◆   The other techniques mentioned in this section can stimulate the brain but PET, MRI and CT scans cannot.

◆   What is it that the PET and fMRI scans can do that the MRI and CT scans cannot do?

◆   The PET and fMRI scans can locate brain activity, whereas the MRI and CT scans can only show structure.

As we said above, each molecule has its own magnetic resonance, which means that magnetic resonance can also be used to locate particular molecules within the living brain. Localized magnetic resonance spectroscopy can be used for the detection of specific molecules including GABA (the neurotransmitter, γ-aminobutyric acid, see Book 4, Section 1.6), glutamine and glutamate in a small volume within the brain.

The techniques described thus far can all be applied to the normal, living brain, inside the skull. Indeed, they have all been used on human participants and have yielded an enormous amount of valuable information about the workings of the brain. This information though falls a long way short of answering all the questions about the brain. Two further avenues of investigation are described below. The first avenue of investigation is of the damaged brain and the second is of the brain after it has been removed from the skull at post-mortem.

2.4.2   The damaged brain

Physical damage to the brain can arise in a number of ways including natural phenomena, medical intervention and scientific investigation. Natural phenomena, for example accidents, ailments, assaults and war, cause damage that is uncontrolled and unique. Thus the extent of the brain damage and the damage to other tissue varies from event to event.

◆   What term describes the scientific status of each of these events?

◆   Each event would be a case report, or anecdote, depending on the reporter concerned.

Medical intervention and scientific investigation, in contrast, are highly controlled and can be repeated with a fair degree of precision on different patients or participants. Medical intervention may involve the removal or cutting of malfunctioning brain tissue, or the removal of a part of the brain to alleviate the symptoms of a disorder. For example, surgery to control severe epilepsy (mentioned in Book 1, Section 1.1.5) or to enable movement in the later stages of Parkinson's disease is not uncommon. Any medical intervention such as these (i.e. brain surgery) has effects on the patient over and above those required to treat the symptoms of the disease. Patients who have had their corpus callosum cut, for example, have great difficulty in naming objects presented on their left, i.e. to their left visual field. The study of these additional effects or side-effects has yielded valuable information about how the brain functions in the absence of the structure removed: the corpus callosum and globus pallidus respectively, in the examples just cited. The side-effects are also reasons for refining or changing the technique.

The process of destroying (or removing) small areas of tissue is called **ablation** and the resulting damage is called a **lesion**. Where a fibre tract or pathway is cut, the procedure is referred to as an '-otomy'; where a structure or part of a structure is removed the procedure is referred to as an '-ectomy'. Thus tracheotomy is cutting of the trachea (windpipe), and appendectomy is removal of the appendix. There are exceptions, however, for example, cutting the corpus callosum is usually known as a commisurectomy, even though it ought to be called commisurotomy, and pallidotomy is the destruction of small areas of the globus pallidus.

An example of what a tumour and a lesion look like on a CT scan is given in Figure 2.11.

Scientific investigation can also involve the removal or cutting of brain tissue, but in this case it might be healthy tissue in an otherwise healthy, non-human subject. Surgery may be with a knife or with a laser that burns away the target tissue. Small areas of tissue can be removed by inserting a cannula and applying suction, a process called aspiration. An even smaller area of tissue can be removed by passing an electric current through a stainless steel microelectrode. The current in this case is stronger than that used to stimulate nerve cells, and the effect is to destroy cells in the vicinity of the electrode tip.

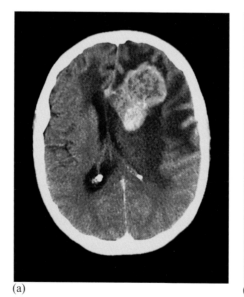 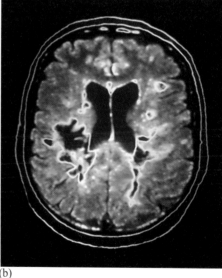

(a)   (b)

Figure 2.11 (a) A CT scan showing a brain tumour and (b) a coloured MRI scan showing the lesions (orange/black areas) caused by multiple sclerosis.

◆   What process would ensure that the electrode tip arrived at the part of the brain intended?

◆   Stereotaxis is used to guide electrodes into the brain.

Chemical ablation is also possible, both of small areas of brain tissue and of specific cell types. A cannula can be used to introduce a chemical (e.g. kainate) that over-excites the cells it contacts to such an extent that the cells die. This technique of **excitotoxicity** is more specific than electrical ablation because only nerve cells are destroyed. Specificity of a different sort is possible with the use of certain chemicals that affect only a particular type of brain cell. One such chemical is 6-hydroxydopamine which is poisonous, but only to those cells that are able to transport it across the cell membrane to the inside of the cell. Dopaminergic and noradrenergic cells do this, so they accumulate 6-hydroxydopamine inside them and are destroyed leaving all other brain cells intact, resulting in a highly selective lesion. 6-hydroxydopamine has been used to investigate the role of dopamine neurons in addiction, a topic considered further in Book 6, Chapter 1. Another drug which targets dopaminergic neurons is MPTP (1-methyl-4-phenyl-1,2,3,6-tetrahydropyridine) which first made its appearance in a batch of illegal designer drugs in California. MPTP causes selective destruction of dopaminergic neurons, but only those of the nigrostriatal pathway in humans and other primates (illustrated in Book 1, Figure 1.19).

Certain techniques prevent a small area of the brain from functioning for a short period of time. The effect is to create a *temporary* or *transient* lesion. One example of this employs a cannula and an anaesthetic. The anaesthetic temporarily inhibits the function of the cells around the tip of the cannula. Another example employs a **cryode**, a device which is designed to carry cold liquids. The cryode is placed on the surface of the brain. Once in position, the cold liquid is passed through the cryode and the cells in a small area around the exposed tip of the cryode cool right down. Whilst they are very cold, the cells are unable to function, but once they warm up, function returns.

◆   How might a lesion be used to yield information about how the brain works?

◆   You could suggest investigating a lesioned brain with *any* of the techniques mentioned in Section 2.4.1.

All of the techniques used to examine the workings of the normal brain can be applied to the damaged brain. For example, those techniques designated 'black box' are useful because the lesion alters the structure of the 'box' and might be expected to alter behaviour. Comparing the response of the normal brain with the lesioned brain gives some insight into the function of the missing (lesioned) area.

2.4.3   The post-mortem brain

A very different kind of examination is possible once the brain has been removed from the skull; it can be visually examined in detail as Einstein's brain was (Book 1, Section 1.1.2). The brain may be sliced to reveal the internal or macro-structure. Different brain regions have a characteristic shape and size, so any abnormalities can be readily observed. Equally, the presence of unusual structures or growths, for example of tumours, and unusual gaps, such as lesions, can be seen.

Such observations must be carried out relatively quickly because dead tissue is likely to decay. It decays partly through the action of its own chemical processes and partly through the action of bacteria. Dead tissue also dries out. These processes of decay and desiccation clearly alter the tissue to be examined, so considerable effort goes in to preserving the tissue and preventing post-mortem decay. Although the preservation process is very good and fast, there remains the risk that some decay occurs. There is also the possibility that the preservation process itself causes some alterations to the tissue being preserved. The most commonly used preservative, or fixative, is formalin, which both prevents decay and hardens the tissue (Figure 2.12).

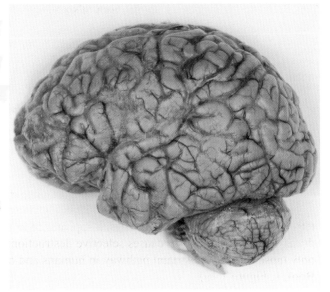

Figure 2.12   Picture of a preserved human brain.

Once the brain is preserved it is possible to cut extremely thin slices, or sections of tissue using a microtome. (The opening sequence of *The Investigative Strand* multimedia package shows a preserved rat brain being sliced with a Vibratome, a microtome with a rapidly vibrating blade.) The sections can be examined under the microscope. The sections need to be between 10 and 20 μm thick (Note: 1000 μm = 1 mm) to see individual cells with a light microscope and less than 0.01 μm thick for use with the electron microscope. The sections are usually stained with a dye (e.g. methylene blue) that attaches to a component (chromosomes in the case of methylene blue) within the cells. The Golgi stain reveals a small percentage of individual neurons in their entirety (Figure 2.13a and also Book1, Figure 2.20b). Horseradish peroxidase is another stain that reveals neurons in their entirety. Stains used with sections for use in the electron microscope usually contain a metal such as lead, uranium or osmium.

The presence and location of any specific chemical in the brain section can be visualized by using the extraordinary specificity of the immune system, by a technique called **immunohistochemistry**. The immune system reacts to the presence of foreign substances, known as antigens, in the body by producing complementary substances called antibodies. The main steps in immunohistochemistry are outlined in Figure 2.14. The antibodies bind with specific antigens, and the antibody–antigen complex is neutralized by white blood cells (which is the normal immune reaction). Appropriate antibodies (e.g. AB) have to be produced and collected for each antigen (e.g. AG), i.e. for each specific chemical, of interest.

Antibodies will bind only to their specific antigens. It is this specificity that is used to determine whether a tissue sample, a section of brain tissue in this case, contains a specific antigen. The brain section is first bathed in fluid containing the antibody, and is then washed. Any antibody remaining on the section after washing must be attached to antigen present in the section. The antibody–antigen complex must now be made visible in some way. This is done by attaching a marker molecule to the antibody. The marker molecule can be fluorescent, so that it can be seen down the microscope, or radioactive so that it can 'expose' photographic paper. An example of the result of using a fluorescent antibody marker is given in Figure 2.15a overleaf. (For clarity, the shape of the neuron is shown in Figure 2.15b.) Either way, the presence of the marker is proof of the presence of the antigen in the tissue. A short video sequence illustrating the immunohistochemistry process features in *The Investigative Strand* multimedia package. In the video, the antigen is Fos, one of the

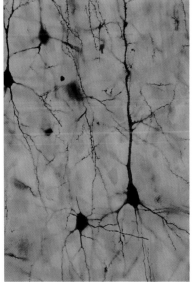

(a)

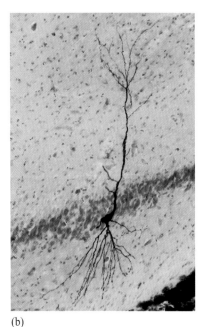

(b)

Figure 2.13   (a) Golgi stained and (b) horseradish peroxidase stained individual neurons.

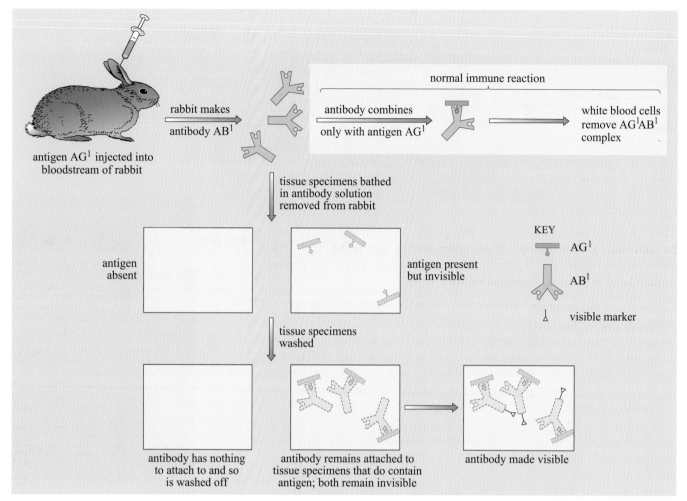

Figure 2.14   Outline of the basic steps in immunohistochemical staining techniques.

first proteins to be produced when a cell is stimulated. Two antibodies are used; the primary, i.e. the first one, attaches to the antigen, Fos, whilst the second attaches to the primary. The complex is made visible when the secondary antibody is made to react with another chemical, producing a yellow colour.

The fact that radioactive substances expose photographic paper is used in another technique to identify where substances are within the brain. This technique is called **autoradiography**. Any substance that accumulates in some cells and not others is a suitable candidate. All that is required is that some constituent, e.g. hydrogen atoms, of the target substance are made radioactive. (Radioactive, or radiolabelled, hydrogen is sometimes called tritiated hydrogen. The substance containing the tritiated hydrogen (^{3}H) is likewise referred to as tritiated, e.g. tritiated thymidine.) One very successful use of this technique has been in the study of the visual system, looking for active cells, i.e. those responding to visual stimulation. One particular type of glucose, called 2-deoxyglucose, 2-DG, is taken up by cells, just like glucose, but it cannot be metabolized, so it accumulates in them. If radiolabelled 2-DG is injected into the bloodstream of an animal exposed to visual stimulation, the 2-DG accumulates in the cells activated by the visual stimulus. Essentially, the more active the cell, the more glucose it uses and the more 2-DG it accumulates. Subsequently, photographic emulsion is put in contact with sections of brain tissue. After a period of time in the dark, the emulsion is processed using normal photographic techniques. The emulsion contains silver salts which react to radioactivity to become silver ions. Only those silver salts immediately next to the radioactive source become silver ions.

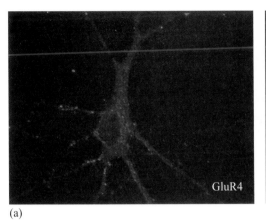

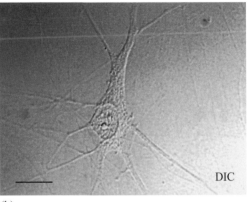

GluR4

DIC

(a)

(b)

Figure 2.15 (a) Fluorescent antibodies attached to glutamate receptor antigens (GluR4) of a single neuron. (b) The same neuron seen under the microscope using differential interference contrast (DIC). The scale bar is 20 μm.

The processing turns the silver ions into grains of silver metal which can be seen under the microscope. Silver grains will be seen in the photographic emulsion in a pattern that exactly corresponds to the pattern of radiolabel in the tissue section (Figure 2.16).

Another way of finding the presence and location of specific substances in the brain that is often used is *in situ* **hybridization**. This technique is used to locate proteins. You will remember from Chapter 1 that genes are turned on to produce proteins. When genes are turned on they produce mRNA (Section 1.5). What is important here is that the intermediate mRNA is a bit like a piece of broken pottery; there is another piece of broken pottery that exactly matches the broken edge. For any one piece of broken pottery, there is only one other piece that will exactly match its broken edge. The same is true for mRNA: for any one piece of mRNA there is only one other piece of RNA that exactly matches. To distinguish the two pieces, or strands, of RNA, the piece of mRNA that codes for protein is called the *coding* strand, and the other piece, that is never naturally present in the cell, the *non-coding* strand. For any piece of mRNA, a radiolabelled non-coding strand can be manufactured in the laboratory. When this radiolabelled non-coding strand is added to a section of tissue it sticks to any matching coding strand present in the tissue. (You might notice a divergence from the pottery analogy here!) The joining together of coding and non-coding strands of mRNA is called hybridization. The tissue section can then be washed.

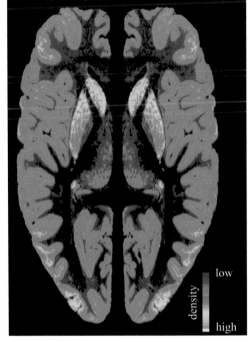

low

density

high

Figure 2.16 Dopamine-D1 receptors in the brain post-mortem, visualized using autoradiography. (The original image was in shades of grey (i.e. black and white) but has been converted into colour for effect.)

◆ By what process would you determine whether there was any radiolabel in the tissue section after washing?

◆ You would use autoradiography.

◆ What would autoradiography tell you?

◆ Autoradiography would tell you not only whether the mRNA of interest was present in the tissue sample, it would also tell you the location of the mRNA of interest in the tissue sample.

◆ How does the presence of mRNA reveal anything about the presence of specific proteins?

◆ The mRNA chosen is that which codes for the protein of interest. If the mRNA is present then it is a reasonable assumption that the protein is too.

The thick, major pathways that link one area of the brain to another can be seen with the unaided eye. However the smaller links and also the links in smaller, non-human brains can only be seen with assistance. A neuron is in a number of respects like a tree: it has branches (axon collaterals) at one end and roots (dendrites) at the other and the two are joined together by a trunk, an axon in the case of neurons. The trunk, axon, is a conduit, with material being moved in both directions, from branches to roots and from roots to branches. Assume for now that the cell body of the neuron is located near the roots. Certain chemicals, called lectins, are taken up by dendrites and transported throughout the neuron. Other chemicals, e.g. fluorigold, are taken up by axon terminals and transported back to the cell body and dendrites. These two processes are called anterograde transport (from cell body to axon terminals) and retrograde transport (from axon terminals to cell body) respectively. These processes can be used to visualize the axon pathways. Either lectins or fluorigold injected into a specific location in the brain will be transported away by the local neurons resulting in the anterograde and retrograde labelling respectively of the neurons. Subsequent examination of brain sections will reveal the axon pathways (Figure 2.17).

Figure 2.17 This image shows the axons (in white) of a brainstem nucleus (the red nucleus) projecting along the brainstem. The image was obtained by injecting the red nucleus in the brainstem with horseradish peroxidase (HRP). Anterograde transport moved the HRP along the axons away from the red nucleus, through the brainstem and towards the spinal cord. Immunohistochemistry localized the places where HRP was to be found. It is not possible to see the red nucleus in this section. The white dots in the cerebellum are retrogradely labelled cell bodies of the interpositus nuclei.

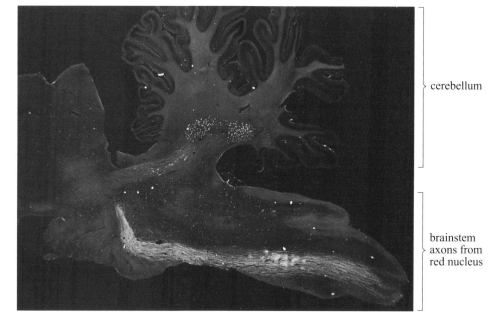

cerebellum

brainstem axons from red nucleus

◆   What technique(s) will allow the presence of lectin or fluorigold in brain sections to be visualized?

◆   The techniques that will allow the presence of a specific chemical to be visualized are immunohistochemistry and autoradiography.

Summary of Section 2.4

There are numerous techniques used for studying the brain. These techniques change as the technology of the day changes. The living and damaged brain must be studied with tools that influence its processes as little as possible. The essence is to observe what the brain does in response to different inputs. The inputs may be natural stimuli delivered through the sense organs, or artificial stimuli, i.e. electricity or chemicals, delivered by electrodes or very fine tubes respectively. What the brain does can be recorded by electrodes, either within or outside the brain (i.e. on the scalp) and again by fine tubes, or by observation of the behaviour of the participant.

Pictures can be taken of the brain in action and areas of high electrical activity or high blood flow can be located. It is assumed that these areas play a substantial role in whatever behaviour any preceding stimulation generates. Damage to the brain may be accidental or deliberate. Both produce lesions and alterations to brain function, but only in the latter case is the lesion precise and specific. The detailed, cellular structure of the brain can only be studied post-mortem. Preparation of uncontaminated slides of brain tissue in which relevant structures are visible requires care, patience and a variety of stains and biochemical techniques. In addition, extremely sensitive methods, such as immunohistochemistry, are able to detect very small quantities of specific molecules. All these numerous techniques provide tantalizing glimpses of the brain at work but none reveal what it thinks.

2.5 Studying animals

The techniques discussed in the previous sections apply equally to humans and to other animals. This section raises some issues that apply only to the latter. (The term animals is used here in the everyday sense to mean all animals except humans.)

Animals are intrinsically interesting, they share a basic biological heritage with people and they are susceptible to many of the same complaints, i.e. diseases and ailments. These three facets of animals are often confused, but shall be considered separately here as they each play an important part in the study of biological psychology.

2.5.1 Animals being themselves

Many people are interested in simply trying to find out what animals do. The number of animals is so vast and their needs and circumstances so different that answering the question 'what do animals do?' will take many years yet. In addition to seeking answers to this question, many people are also seeking answers to the question 'why do animals do what they do?'. *Causal* answers to the 'why' question (actually the same answers as to the question 'How do animals do that?') are fascinating: how do flies learn, toads find a mate, eels find the Sargasso sea? Unfortunately, the answers to these questions do not feature in this course, unless those answers inform the primary (i.e. human) content. The point being made here is simply that the primary focus of some research is animal behaviour; and for this type of research, any connection subsequently made with human behaviour is entirely fortuitous.

2.5.2 A shared biological heritage

Sensible connections can be made between animal behaviour and human behaviour because of their shared biological heritage. The basic building blocks of all animals are the same. These building blocks, cells and their contents, were considered in Book 1, Chapter 2 and in Chapter 1 of this book. The following examples have simply been chosen to reinforce the point. Exactly the same molecule, dopamine, in exactly the same area of the brain, the substantia nigra, is involved in the control of movement in both the rat and the human (see Book 1, Figure 1.19). Exactly the same molecule, cAMP, is involved in learning in the mollusc *Aplysia californica*, the fruit fly *Drosophila melanogaster* and the rat *Rattus norvegicus*. This concordance is a consequence of evolution and informs the discussion about the 'whys' and 'wherefores' of human behaviour. As a consequence of this concordance, the very early stages of development can be studied in the African clawed toad *Xenopus laevis* and in the chick *Gallus domesticus*, for example, with a view to revealing secrets of development at the cellular level that apply to all animals. Thus many animals are studied because of the information they will reveal about animals in general.

2.5.3 Common complaints

It is not just the basic building blocks that people share with other animals, they share diseases too. Narcolepsy is one such common complaint: it occurs naturally in certain breeds of dog, including poodles, dachshunds, Labradors and Dobermann pinschers, as well as in people. Narcolepsy is a disabling neurological disorder that affects the control of sleep and wakefulness. It may be described as an intrusion of the dreaming state of sleep (called REM or rapid eye movement sleep) into the waking state. Unrelenting excessive sleepiness is the most prominent symptom of narcolepsy. Patients with the disorder experience irresistible sleep attacks, throughout the day, which can last for 30 seconds to more than 30 minutes, regardless of the amount or quality of prior night time sleep. These attacks result in episodes of sleep at work and social events, while eating, talking and driving, and on other similarly inappropriate occasions. Symptoms generally begin between the ages of 15 and 30 years. The four classic symptoms of the disorder are excessive daytime sleepiness; cataplexy (sudden, brief episodes of muscle weakness or paralysis brought on by strong emotions such as laughter, anger, surprise or anticipation); sleep paralysis (paralysis upon falling asleep or waking up); and hypnagogic hallucinations (vivid dream-like images that occur at sleep onset). Disturbed night time sleep, including tossing and turning in bed, leg jerks, nightmares, and frequent awakenings, may also occur. The development, number and severity of symptoms vary widely among individuals with the disorder. Although narcolepsy is not a rare disorder, it occurs in about 1 in 2000–3000 people, it is often misdiagnosed or diagnosed only years after symptoms first appear.

◆   Which two of the four symptoms of narcolepsy do you think would be evident in dogs?

◆   The first two: excessive daytime sleepiness and cataplexy. Sleep paralysis and hypnagogic hallucinations are very difficult to diagnose in dogs.

Excessive daytime sleepiness is also difficult to evaluate in animals (and impossible in cats!) because of their special sleep/wake patterns. Dogs do not have a solitary, extensive period of wakefulness during the day and will nap several times during their active period, so the diagnosis of excessive daytime sleepiness is difficult to make. However, in an unusual twist, studies on humans have aided the diagnosis of narcolepsy in dogs. Human sleep is often measured with EEG recordings, and it is possible to adapt EEG techniques for use with dogs. Twenty-four hour EEG recordings have shown that the sleep patterns of narcoleptic dogs are very similar to those of narcoleptic people, except that the sleep is more fragmentary and the onset of REM sleep is much quicker.

Cataplexy has none of the drawbacks of the other three diagnostic criteria; it is conspicuous and unmistakable. Cataplexy occurs when the afflicted dogs become excited, e.g. when offered their favourite food or by vigorous play. The dogs suddenly and temporarily lose skeletal muscle tone, mostly of the back legs and neck. The cataplectic attacks last a few seconds during which the dogs remain awake and fully conscious (Figure 2.18).

Adult dogs of the same breed and comparable ages show enormous variation in their symptoms. Some 'show dogs' will consistently have cataplexy regardless of the circumstances or audience, whereas others, including littermates will have no symptoms at all. Our understanding of cataplexy, and as a consequence, narcolepsy, was advanced, quite by luck, by unrelated studies on appetite.

12 seconds

13 seconds

24 seconds

50 seconds

Figure 2.18 Stills taken from a movie in which a cataplectic attack occurred in two dogs. The times below the images indicate the time from the start of the video sequence. The cataplectic attacks seen here (at around 24 seconds) lasted about 23 seconds.

Two separate groups of researchers were looking for chemicals in the hypothalamus. They published the results of their searches within a month of each other in 1998, one in the journal *Proceedings of the National Academy of Sciences* and the other in *Cell.* (The references can be found in Siegel *et al.*, 2001.) One group (Sutcliffe's group) named the molecules they had found hypocretins and the other group (Yanagisawa's group) named the molecules they had found orexins. It turned out that the two groups had found the same molecules, hereafter referred to as hypocretins.

Yanagisawa's group investigated the role of the hypocretins in appetite by producing a mouse (*Mus musculus*) that could not produce them. Using the techniques of genetic engineering it is possible to remove, inactivate or add specific genes to organisms. Where a specific gene is removed (or made non-functional) from an organism, that organism is said to be a **knockout** organism (see Box 2.1). Where a specific allele is added to an organism, that organism is said to be a **transgenic** organism. Yanagisawa produced a strain of mice that have brain cells that cannot produce hypocretin, and without hypocretin these mice cannot send the normal hypocretin message. The knockout mouse had a reduced appetite as expected but not dramatically so, and the absence of hypocretin had no effect on body weight. In a good example of broad based research, Yanagisawa's group decided to watch the mice and observe their feeding behaviour, rather than just weigh the amount of food left at the end of the day. What they observed were occasional, sudden, temporary movement failures; the mice were cataleptic (i.e. in a state where the body becomes stiff and stops moving). The mice may actually have been cataplectic because further studies using EEG recording confirmed that the mice were also narcoleptic.

Whilst it would be technically possible to delete specific genes from people, i.e. to create 'knockout' people, the issue is subject to considerable ethical debate. And if the sole reason for creating a knockout person was for scientific investigation, the procedure would be deemed unethical. Scientifically, it would also be a fairly daft thing to do. Any outcome would not be known for ten years or so and, for reasons explained in Section 2.3, the outcome may be very inconclusive. The knockout mouse, in contrast can yield substantive information in a relatively short period of time.

There are several qualities that make the mouse a suitable organism for scientific studies. These include the short life cycle, rapid maturation, multiple offspring, well known and documented genetics, ease and economy of housing and maintenance.

There are of course many other organisms which have these qualities, not least is the fruit fly *Drosophila melanogaster* (Figure 2.19a) and the tiny nematode worm *Caenorhabditis elegans* (Figure 2.19b). According to some authorities, up to 65% of human genetic diseases have some counterpart in this worm.

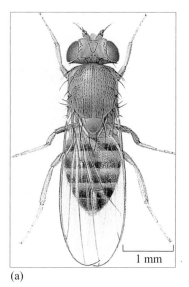

(a)

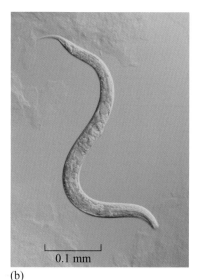

(b)

(c)

Figure 2.19 (a) *Drosophila melanogaster*; (b) *Caenorhabditis elegans*; (c) *Mus musculus*.

Box 2.1 Knocking genes about

Knockout organisms lack a gene of interest throughout their entire lives and throughout their entire bodies. More subtle techniques add biochemical switches to genes, allowing them to be switched on or off at appointed times and in appointed tissues. One technique uses a dietary component to act as a switch. If the component is present, the gene functions normally; if the component is absent the gene is switched off and it cannot produce its protein. This is a 'knockdown' organism.

Alleles can be added to an organism using a virus, which acts as a vector. The chosen virus is able to infect cells and exert some control over transcription, effectively reversing the transcription process, causing the cell to insert a new piece of DNA into the genome. If the virus has been engineered so that the inserted piece of DNA is not harmful, but rather is beneficial, intact alleles can be placed in genomes which contain damaged alleles. This is viral transfection. (Some children with X-linked, severe combined immunodeficiency syndrome have received this type of treatment.)

All these techniques have effects, often unforeseen, over and above those intended. Side-effects are not uncommon in treatments of various kinds, but if the treatment alters the genome, treating the side-effect becomes highly problematic.

Both the narcoleptic dog and narcoleptic mouse could be referred to as **animal models** of human narcolepsy, meaning simply that there are similarities between their conditions. Note that the term animal model does not mean engineered specifically for the purpose of scientific investigation, but covers both the natural and the engineered animal disease state, dog and mouse narcolepsy respectively. Where there is an animal model, the genetics and environment of the organism can be controlled. Furthermore, the timing between key events such as a behaviour and an examination of the microstructure of the brain can be controlled.

◆   Why might the features of animal models just mentioned be important?

◆   Those features are important because they increase the likelihood of being able to demonstrate a relationship between cause and effect.

2.6 Tissue culture

Small quantities of tissues and cells, kept alive and functioning outside the body, are called tissue or **cell cultures**. They are outside the body and so are called *in vitro*, as opposed to cells or tissues inside the body which are *in vivo*. Provided the fluid surrounding the cells has the appropriate composition and nutrients, cells can survive and grow, sometimes for considerable periods of time. Some cells, often known by abbreviations, such as HeLa, or COS cells, have been replicating in culture for many years. It is estimated that there are two tons of HeLa cells in various laboratories around the world. These, so-called immortalized cell lines, often originate from tumours. Cell cultures can be very useful where the object of investigation is to understand cellular or molecular mechanisms.

However, isolated cells tend not to maintain their usual structure. (This is a bit like taking off tight clothing; our 'shape' is maintained when we are confined, but we tend to lose shape and 'spread' a little once out of the restrictions of specific environments.) This applies both to individual cells and to groups of cells (which may also lose the usual orientation they might have had with respect to each other). This drawback of isolated or separated cell cultures can, at least in part, be resolved by the use of organotypic cultures.

Organotypic cultures are, as they sound, cultures of whole or parts of whole organs. Organotypic cultures tend to maintain most of their cellular contacts and so can be useful for studies examining local cell interactions. Like cell cultures from adult or normal tissue, they have a finite lifespan (from days to weeks). This means that, unlike immortalized cell lines, each culture can only be used for this finite period of time. If experiments are to be repeated, further cultures are required. It is not possible to propagate cultured organs, because of their finite lifespan, and so replication of results can be complex.

Brain slices can be maintained in this way. Figure 2.20 shows some of the possibilities of maintaining cells *in vitro*. A slice of brain tissue from the rat, 200 μm thick, was examined after fourteen days *in vitro*. The response of a specific cell (a Purkinje cell from the cerebellum) to three different sorts of stimulation was examined. The experiment had been set up so that any change in calcium concentration anywhere in the cell would cause that place to fluoresce. (The change in calcium concentration is a measure of the cells activity; the greater the change, the greater the activity.) The Purkinje cell of interest was stimulated via synaptic activity from three other neurons. Clearly, stimulation by neuron 1 had a far greater effect on the Purkinje cell than stimulation by neurons 2 or 3.

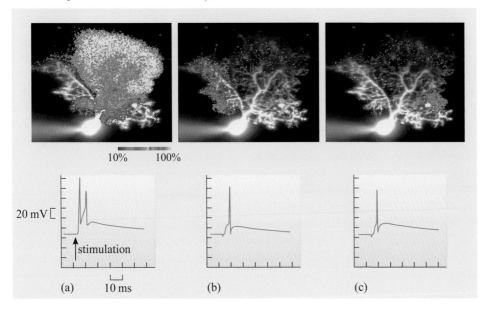

Figure 2.20   The response of a single Purkinje cell to stimulation from three different neurons. (a) Response of the cell to stimulation by neuron 1, (b) response of the cell to stimulation by neuron 2, (c) response of the cell to stimulation by neuron 3.

Cell culture, though, is not without problems; two are considered here. Firstly, the original source of the cells raises some particular concerns. Cells may be taken from embryonic or fetal tissue, as such cells tend to survive and grow better in culture than cells taken from adult tissues. Some fetal cells, known as multipotent stem cells, are able to differentiate into, i.e. become, any one of a number of different kinds of cell, such as glia, neurons, or liver cells. (Differentiation is considered in Chapter 3.) The use of embryonic or fetal cells raises ethical issues, not the least of which is who can give consent to their use? On the other hand, immortalized cell lines derived from tumours (cancers) do not raise ethical issues but may differ in some crucial way from non-tumour cells. Secondly, cells in a body function as part of an integrated system. The body has very finely controlled mechanisms for monitoring and adjusting the environment in which cells operate. The control mechanisms are absent in cell culture which means that the functioning of the cell in culture may differ from the functioning of the cell *in situ*, i.e. in its normal location.

Summary Sections 2.5 and 2.6

Animals other than humans provide insights into how brains work. In part this is because they share, to differing extents, common biological processes. They also share common complaints. Investigations into such complaints can be pursued using knockout and transgenic organisms, options that are not available in human studies. It is also possible to investigate some cellular mechanisms using cell culture. Caution is needed in both cases in extrapolating from any results: just as animals are not people, so cells are not organisms.

2.7   Participants in experiments

The procedures described in Section 2.4 are performed on animals, both human and non-human. This short section outlines the rules that govern what sorts of experiments may be undertaken and how the participants should be treated.

2.7.1   Animal experiments

Our understanding of brain and behaviour has been considerably advanced by knowledge gained from experiments with animals. This fact raises two separate issues. The first issue is about whether such experiments should be conducted and if so, under what circumstances. No country has banned animal experiments. Hence, at the time of writing (2003), all law-making bodies have decided that the humanitarian advantages of animal experiments outweigh other considerations. Those who hold the view that animal experiments should not be allowed have not yet persuaded the law-makers of the merits of their case. They have, however, influenced the laws and codes of practice governing the use of animals in experiments. The United Kingdom Home Office regulations on the Use of Animals in Experiments (the Animals (Scientific Procedures) Act of 1986) control what experiments can be done in the UK. They also specify the minimum conditions in which animals may be kept. The regulations are changed from time to time by the Home Secretary who regularly issues updates in various 'codes of practice' documents. Other countries have their own regulations, some of which are more stringent than the UK, and some less so. (The focus here on the regulations of the UK, reflects the base of the Course Team and has no other implications.) The regulations are strictly enforced by a series of inspectors and local veterinarians. In addition, most institutions where animal work might be conducted have local ethical committees, which examine in detail every proposal to use animals and can veto this work if they are not satisfied that it is justified, useful and 'humane'.

The second issue raised by animal experiments is how to use the information those experiments generate. This course would not be possible without some of that information; the nature of the subject matter of this course requires that the results of some experiments on animals are reported. But that does not mean that *all* information generated in animal experiments is equally worthwhile. Some experiments, especially those conducted many years ago, would not now be permissible in the UK. So should information gained from such experiments be reported? There are many facets to this issue, but two were used as guiding principles in writing this course. The first was how important the information was deemed to be; the more important the information was deemed to be, the less assiduously the second principle was applied. The second principle was to report, as a first choice, experiments that meet current (2003) Home Office regulations.

2.7.2 Human participants

The guiding principle on the use of people in medical and scientific experiments is that they should have sufficient information about the experiment and the consequences of participation to decide for themselves whether to participate. This is the principle of **informed consent**. People who have given informed consent are considered to be partners in the experimental process and are referred to as participants. The term participants has superseded the term subjects, the latter having connotations of powerlessness, a state that is in conflict with informed consent. (Animals do not give informed consent and may still be referred to as subjects, meaning the focus of attention.) Unfortunately, the application of the principle of informed consent has varied from country to country. To rectify this, in October 2000 the Declaration of Helsinki stated that the same ethical rules should apply to all human research, irrespective of location.

2.8 Summary of Chapter 2

The function of this chapter was to introduce you to the scientific process as it applies to biological psychology. Some theoretical, practical and ethical limitations on the study of the brain were considered. A number of techniques were described, both because of their importance and because they illustrate the scale of the task.

The brain can be examined in a number of different ways. The techniques applied differ, depending on whether the brain is living or dead. Investigating the living brain usually requires a stimulus, which may or may not be naturally given, and a response, which again may or may not be naturally observed. Some of the techniques are invasive and some are non-invasive.

The use of animals and of tissue culture in gaining insights into the functioning of the brain was also considered.

As you progress through the course you should refer back to these techniques as they are mentioned and elaborated upon.

Learning outcomes for Chapter 2

After studying this chapter you should be able to:

2.1 Recognize definitions and applications of each of the terms printed in **bold** in the text.

2.2 Recognize and understand the uses and limitations of a variety of techniques used to investigate the brain.

2.3 Distinguish between techniques that are invasive and non-invasive; suitable for living and non-living brains.

2.4 Describe the principles of the EEG, CT scans, PET scans, MRI scans, autoradiography and immunohistochemistry.

2.5 Be aware of possible theoretical, practical or ethical constraints to particular experiments.

2.6 Explain the ways in which insight into biological psychology has been gained by studying animals.

Questions for Chapter 2

Question 2.1   *(Learning outcome 2.2)*

To locate a suspected brain tumour (lump of unwanted tissue), would you use an EEG, microiontophoresis or a CT scan? Explain your reasons.

Question 2.2   *(Learning outcome 2.3)*

Here are five techniques which can be used for studying some aspect of the brain:

(i) EEG, (ii) DBS, (iii) retrograde labelling, (iv) MEG, (v) *in situ* hybridization.

(a) Which one of them requires only post-mortem brain tissue?

(b) Which of the four remaining techniques are invasive and which non-invasive?

Question 2.3   *(Learning outcome 2.4)*

State whether autoradiography and immunohistochemistry require:

(a) samples of tissue

(b) antibodies

(c) radioisotopes (radioactive material)

Question 2.4   *(Learning outcome 2.4)*

Although the EEG is old technology it has one major advantage over all the more recent scanning and imaging techniques; what is that advantage?

Question 2.5   *(Learning outcome 2.5)*

Experiments involving the effects of alcohol on motor performance have been suggested in the past as possible home experiments for Open University students. Why do you think such suggestions have invariably been rejected?

Question 2.6   *(Learning outcome 2.6)*

What specific information was identified as having been gained from studying an animal model of narcolepsy?

THE MAKING OF INDIVIDUAL DIFFERENCES

3.1 Introduction

This chapter addresses the question of how the differences between individuals, especially in behaviour, arise during development. Development, the transformation of the single cell, the zygote, into an adult organism with billions of cells, numerous organs and an intricate, functioning nervous system, is one of the most remarkable feats of living systems. The process begins when an egg cell, or ovum, is fertilized by a sperm, or spermatozoon. The resultant single cell, the zygote, divides to produce two cells and each of these divides again to produce four cells; and division continues apace producing the thousands and eventually the billions of cells of the adult organism. The cells grow, interact, move around and diversify under the guidance of the genetic material each cell contains, producing different structures and features and the characteristic shape of the organism. The sequence of events that occurs during that period of rapid change that begins at conception and continues through to the time when the relative stability of maturity is attained, i.e. development, has been described in considerable detail for many organisms. Some features of the formation of the early nervous system can be seen in the 'Brain' section of the multimedia package *Exploring the Brain*, and the shape and size of the human brain at a number of different ages is presented in Section 3.2, but the sequence of events will not be the main focus of this chapter. Instead, the chapter focuses on individual differences and, in particular, on how those individual differences arise.

Book 1, Sections 1.4.5 and 1.4.6 identified the two key causes of individual differences between organisms. The first cause was stated as *differences in genes between people can manifest in differences in their nervous systems and thereby contribute towards differences in their behaviour*. The way in which genes affect behaviour, or the way differences in genes translate into differences in behaviour, is a recurring theme throughout this chapter, but is considered explicitly in Sections 3.1.1 and 3.8.

The second cause of individual differences was stated in Book 1, Section 1.4.5 as *the developing body acting in dynamic interaction with the environment*. There are three consequences of this dynamic interaction, which lead to three different kinds of question, and they in their turn lead to three different kinds of investigation. The dynamic interaction is continuous and the environment constantly changing but the situation can be simplified by considering a single, identifiable environmental event. The three questions then become:

1   Is the behaviour of the organism affected at the time of the event?

2   Is the behaviour of the organism affected at some later time, after the event?

3   Through what mechanism is the interaction, between the body and the event, mediated?

These questions can be represented on a schematic diagram of development (Figure 3.1).

Figure 3.1 Schematic representation of five stages of development. The length of each stage represents the amount of change that occurs during that stage, not the duration of each stage. The circled numbers represent questions 1, 2 and 3 in the text.

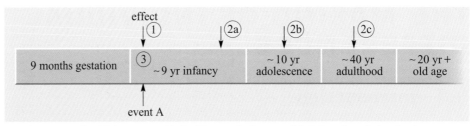

So, for example, if the event (A in Figure 3.1) were high levels of physical contact during infancy (i.e. lots of cuddling), then this might affect (a) the behaviour of the infant at the time (effect 1 in Figure 3.1) and (b) the behaviour of the child/adolescent/adult years later (effects 2a, 2b, 2c respectively in Figure 3.1). If there is an effect, then there is a mechanism by which physical contact caused those effects (3 in Figure 3.1), and this mechanism can be sought. Note that in Figure 3.1, the length given to each stage of development represents the amount of change that occurs during that stage, not the duration of each stage.

The first part of this chapter (Sections 3.3 and 3.4) considers questions 1 and 2, which can loosely be regarded as the phenomenology that is the conspicuous consequences of developmental events. The second part of the chapter (Sections 3.5, 3.6 and 3.7) considers question 3, the mechanisms by which events during development exert their effects. The final part of this chapter, Sections 3.8 and 3.9, returns to the issue of genes and their influence on development, the nervous system and behaviour. To finish this introduction, the problem of relating genes to behaviour is considered more fully.

3.1.1   The 'genes and behaviour' problem

Amidst the progressive change to the brain and nervous system that occurs during development, there is one constant, one fixed element; the set of deoxyribonucleic acid, DNA, molecules found in each cell. These molecules are the genetic material, and they store the information for the construction of the organism. The same set of DNA molecules is found in every cell of an organism (its genome). (There are some exceptions to this such as sperm, unfertilized egg and red blood cells, but they need not concern us at present.) A different set of DNA molecules is found from one organism to another. Each person has a slightly different set of DNA molecules from each other person, i.e. there are genetic differences between people.

The chain of events that links the genome to protein production has been worked out in some detail, and the mechanism is universally accepted and very similar across all organisms. (See Chapter 1 in this book.) What has proved problematic is extending that chain of events from the genome to behaviour; how the genome, a set of molecules that guide protein production, affects behaviour, several levels of organization away from molecules, is not easy to visualize. Take, for example, the garden spider (*Araneus diadematus*). It hatches from an egg sac sometime in the spring, wanders about for a while and then constructs an orb (disc-shaped) web; a web characteristic of the species (Figure 3.2a). A spider of a different species, say the triangle spider (*Hyptiotes paradoxus*), builds a triangular web (Figure 3.2b), whilst the wolf spider (*Pardosa proxima*), builds no web at all. All these spiders produce silk, yet what they do with it varies from species to species, in a species-specific way. Spiders receive no outside instruction in silk use (i.e. they do not learn from other spiders or books!). What is puzzling is the fact that the different species use their silk in different ways. Why doesn't the wolf spider build an orb web? Why doesn't the garden spider build a triangular web?

The only information every spider can be guaranteed to have is in its genome. The genome varies from species to species, and it follows that the genome *must be sufficient* to ensure that the spiders use their silk in a way that is characteristic of the species. So, how does the genome affect web design, such that an orb web, a triangular web or no web is constructed? That the genome influences the structures and musculature of the spider by producing or not producing certain proteins at particular times is just about possible to imagine, though by no means easy. It is also possible to argue that the size, shape and strength of the spider's legs and body have a direct effect on the distances between threads and tension of a web. This is fine if the only problem is differences between orb webs. But that is not the only problem. The real problem is how genes can affect web design, to produce an orb or a triangle, and how genes can affect whether silk is used to construct webs at all. The real problem is how differences in the genome cause the differences in behaviour between the three species of spider. At the moment there is no satisfactory resolution to this problem.

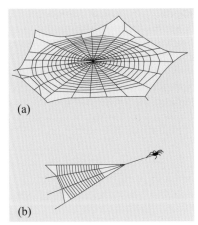

Figure 3.2 Webs constructed by (a) the garden spider (*Araneus diadematus*) and (b) the triangle spider (*Hyptiotes paradoxus*).

Now think about people and the range of behaviours they exhibit. How do genes influence human behaviour? The problem is even more intractable than with the simple spider example. Note that this discussion has accepted that genes affect behaviour; logically, in the case of the spiders, they must do. The difficulty is in explaining *how* genes affect behaviour. It is very difficult to conceptualize. So in the following discussion, when differences in genes are linked to differences in behaviour, remember that there is not a neat, sequential, causal relationship between a gene and a behaviour; it's more complex than that.

The chapter returns to this discussion of genes and what they do, in Sections 3.8 and 3.9. The next section describes the changes in the nervous system during the early months of life.

3.2 Growth and development: the big picture

The scale of the problem facing the human zygote is vast. The zygote, the single cell resulting from the fusion of a sperm and an ovum, is about the size of a full stop on this page, yet within 9 months it has become a 3 kg, 50 cm long baby. The single cell has made millions of other cells; 10^{11} (i.e. 100 000 000 000) neurons in the brain alone. Not only is there this vast and rapid increase in the number of cells (some calculations put the peak figure at 250 000 new neurons being born every minute!), but there is also the problem of organizing the constantly growing population of cells into organs and tissues. For neurons there is the additional problem of making appropriate connections between each other, to effector organs (e.g. muscle) and with sensory organs.

The mechanisms which control and direct the development of the nervous system are, for the most part, beyond the scope of this course. The growth of axons, though, is considered in some detail in Section 3.6.1. Here, a brief account of what happens as the nervous system develops is given to provide an overview of what must be one of the most remarkable processes on Earth.

The zygote divides in two and those two cells also divide in two. The resulting four cells divide in two, producing eight cells and so on. The process of division continues with very little increase in the size of the sphere of cells until day seven when implantation into the uterus occurs. By embryonic day 15 (E15) the various structures necessary to sustain the embryo are in place. These structures, which

Figure 3.3   The first embryonic structure to form in the germ disc is the primitive streak, which appears on day 15. (a) Section through the embryo showing the location of the germ disc between the amniotic cavity and the yolk sac. (b) A more detailed view of the germ disc showing the primitive streak forming in the centre.

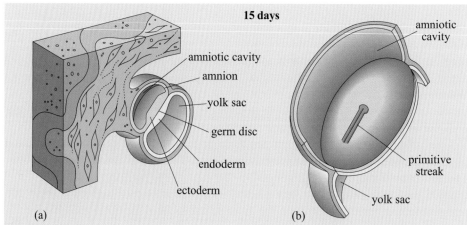

15 days

amniotic cavity

amnion

yolk sac

germ disc

endoderm

ectoderm

(a)

amniotic cavity

primitive streak

yolk sac

(b)

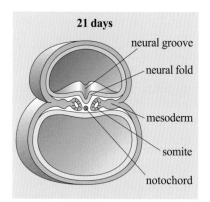

21 days

neural groove

neural fold

mesoderm

somite

notochord

Figure 3.4   Diagrammatic cross-section of the embryo (at 21 days), showing the early stages of neural tube formation above the notochord.

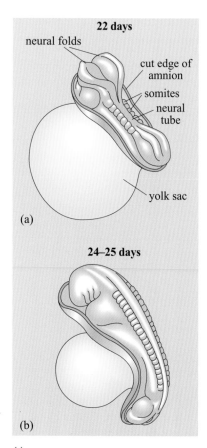

22 days

neural folds

cut edge of amnion

somites

neural tube

yolk sac

(a)

24–25 days

(b)

include the placenta, amniotic cavity and yolk sac, have all been produced by the embryo through the processes of cell differentiation (the transformation of one cell type into another) and cell migration (the movement of cells from one place to another). At this stage, the embryo itself is a flat **germ disc** comprising two layers of cells: a layer of endodermal cells with the flattened sphere of the yolk sac to one side and a layer of ectodermal cells with the flattened sphere of the amniotic cavity to the other side (Figure 3.3a). This thin little germ disc is the ultimate source of all the cells in the human body.

On about day E15, the two layers of cells that comprise the germ disc begin to separate from each other forming a cavity between them. Ectodermal cells flow into the cavity creating a third layer of cells called mesoderm between the endoderm and ectoderm. The movement also creates a groove called the primitive streak along the midline of the germ disc (Figure 3.3b). This is the first sign of any structure within the embryo itself. The primitive streak defines the central axis of the future human being, where the backbone will form. The three germ layers (ectoderm, mesoderm and endoderm) produce and receive chemical signals causing further changes in structure. The cells that occupy the central axis of the mesoderm exhibit increased adhesion and stick tightly together, separating from their neighbours and producing a rod shaped structure called the **notochord**, which is the beginning of the backbone. Now the cells on either side of the notochord become adhesive and they stick together, but not enough to cause complete separation from their neighbours. These clumps of cells are somites which will develop into the muscles that lie on either side of the backbone. At the same time as the somites are being formed, the ectodermal cells on top bulge up on either side of the midline, and these neural folds form a neural groove, shown in cross-section in Figure 3.4.

The upper part of the neural folds come together and fuse along the midline of the embryo producing the **neural tube**, from which the brain and spinal cord develop (Figure 3.5). There is a sequence on the development of the brain in the multimedia package *Exploring the Brain*.

The cells that form the neural tube will eventually give rise to neurons (they are neural precursor cells). Whilst they retain the ability to divide they are **stem cells**, but once they stop dividing they are **neuroblasts**. Those neuroblasts near the notochord will become motor neurons, whilst those further from the notochord will become

Figure 3.5   The neural folds first meet and fuse on day E21 to produce the neural tube in the middle of the embryo by day E22 (a). Fusion then proceeds towards the anterior end, where the brain will form and the posterior end where the spinal cord develops (b), with growth and elongation resulting in the formation of a curling tail.

sensory neurons. The cells just above the neural tube form the **neural crest** and these cells migrate away from the neural tube to form the neurons and glia of the sensory and sympathetic ganglia, the neurosecretory cells of the adrenal gland and the enteric nervous system, depending on which migratory path the cells take. (See Figure 3.6.)

The neural tube goes on to form the brain and spinal cord. The tube swells at one end, this will become the forebrain, and there are two smaller swellings behind it, which are the rudiments of the midbrain and hindbrain. At these swellings the hollow core of the tube becomes the ventricles. Within the walls of the tube surrounding the ventricles, stem cells divide to produce neurons and glial cells, which move and adhere to form the characteristic structures of the brain, such as the cerebellum, medulla, etc.; neurons clump together to form distinct structures and axons begin to link them together. At five or six weeks the tube has to bend to accommodate its own growth within the skull. Two bends occur at very specific places and thereby locate the brain in its characteristic position with respect to the spinal cord. From then on there is a rapid appearance of recognizable structures.

The remarkable similarity in the early stages of development between a variety of different vertebrates was first illustrated by Haeckel in 1874. His drawings of embryos are still to be found in many texts (e.g. Figure 3.7a), but are now regarded as overemphasizing the similarities.

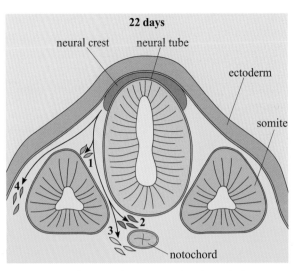

Figure 3.6 Diagram of a cross-section through the developing mammalian embryo at E22. The neural crest cells follow four distinct migratory paths. Cells that follow path 1 become sensory ganglia; cells that follow path 2 become sympathetic ganglia; cells that follow path 3 will become adrenal neurosecretory cells; cells that follow path 4 become enteric nervous tissue, serving the gut and also non-neural tissues such as cartilage and pigment cells.

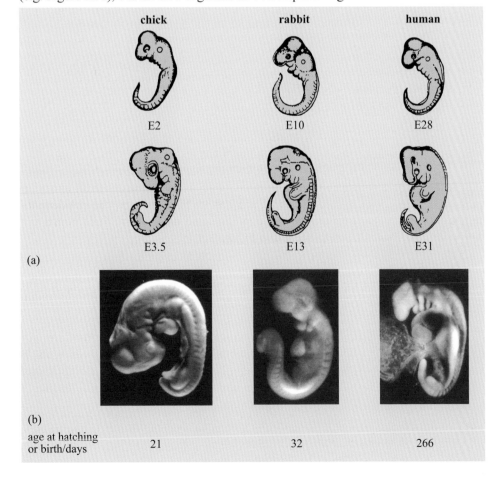

Figure 3.7 (a) Original drawings of three embryos at two stages of development by Haeckel.
(b) Photographs of the three embryos at the tail bud stage. The embryonic ages shown are approximate.

The similarity is evident both at a structural and a biochemical level. A similarity at the biochemical level between different species is not unusual; a number of other processes, e.g. respiration and neurotransmission share the same biochemical pathways, and the similarity extends beyond vertebrates, to flies and even worms in some cases. The term used to describe the situation where a particular biochemical pathway is the same or has been conserved across species is biochemical parsimony.

Once the human embryo has developed a human body plan, at about E56, it is, by convention, called a fetus, also 56 days old. By the end of the third month after conception, cerebral and cerebellar hemispheres are obvious, and the thalamus, hypothalamus and other nuclei within the brain can be distinguished. In the following month the cerebral hemispheres swell and extend. By the fifth month the characteristic wrinkles of the cerebral hemispheres begin to appear. Most of the sulci and gyri are apparent by the eighth month of development although frontal and temporal lobes are still small by comparison with the adult, and the total surface area is much below its eventual size. (See Figure 3.8.)

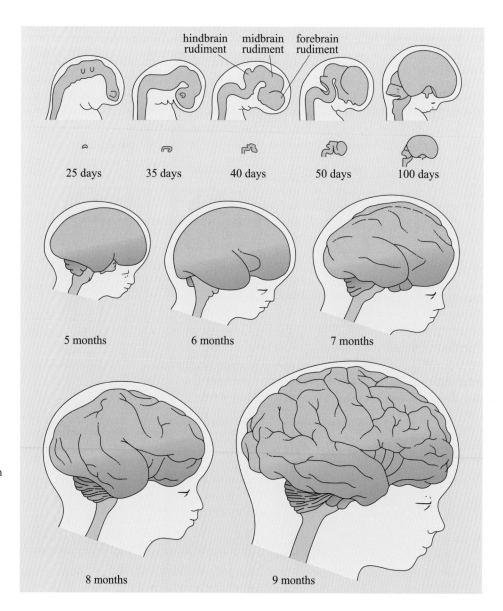

Figure 3.8   The growth of the human brain. The diagrams in this sequence are about three-fifths life-size. The first row shows the brains of the second row enlarged for clarity. Note the exceptional growth of the part of the brain which gives rise to the cerebral hemispheres (labelled as the forebrain rudiment at 40 days).

Virtually the full adult complement of neurons in the human brain is present shortly after birth, though the extent to which stem cells continue to produce neurons throughout life is not known. Glial cells continue to increase in number until adolescence, when the mature adult brain structure is achieved. It is largely the increase in glial cell number and the growth of axons, myelin and dendrites that leads the 350 g newborn brain to become the 1300 g adult brain.

There are many developmental paths to maturity and beyond, so you should not think of development as proceeding in an inevitable way along a prescribed path. As the organism grows and develops and acquires a history, so its attributes of size and shape and abilities, its **phenotype**, serves to distinguish it more and more from all other organisms. And although changes to the phenotype are continuous, they become more subtle as time proceeds: the embryo creates pathways and brain structures, the adult modifies connections and synapses (Figure 3.1).

The extent to which various environmental factors impact upon development is considered in the next section.

3.3 The unique phenotype

We each begin life with a unique genome. As we grow and develop, we are each subjected to a range of factors that influence the way development proceeds. Most of those factors are common to us all, the intracellular and intercellular signals, hormones, birth, milk. But the precise combination and the range and duration of those factors varies between individuals, such as the duration of gestation or the composition and quantity of a mother's milk, for example. In addition we each undergo different experiences, injuries and illnesses. What and who we are is our phenotype and it depends on all these factors. And because no two people have all these factors in common, each person (each phenotype) is unique.

Let us now consider in a little more detail some of those factors and how they influence our phenotype.

3.3.1 Small babies

Development continues in the womb until birth, which, in humans, is about 38 weeks after conception. (The often quoted duration of pregnancy of 40 weeks is based on pregnancy beginning on the first day of the last menstrual period.) The duration of the period of development before birth, called gestation, is highly variable. It is not possible to determine its full range in the UK, because medical intervention usually prevents pregnancies going beyond about two weeks after the due date, or term, although 10 month (43–44 week) pregnancies are not uncommon. Many babies, however, are born well before their due date. Those born more than three weeks before their due date are called pre-term. Pre-term babies clearly have a very different experience from full-term babies.

◆ Consider two babies, 38 weeks after conception, one of them is pre-term, born say at 35 weeks. In what ways does the experience of the 38-week-old pre-term baby (born at 35 weeks) differ from that of the full-term baby?

◆ There are many differences that could be mentioned. For three weeks, the pre-term baby has been physiologically independent; directly exposed to light, smells and air; able to stretch (no longer constrained within the womb); in direct physical contact with objects, (no longer buffered by the mother and surrounding amniotic fluid).

It is reasonable to ask whether these differences, which may last for eight weeks or more depending on how many weeks pre-term the baby is born, affect the development of the baby.

Consider the data in Table 3.1. The figures are means with the range of values in parentheses. In this study, full term was 40 weeks.

Table 3.1   Comparison between 25 pre-term and 39 full-term babies on two measures taken at birth (neonatal measures) and three measures taken in their 9th year (i.e. after their 8th birthday). (Full term was 40 weeks.)

Measure	Pre-term ($N=25$)	Full-term ($N=39$)	Probability values
Neonatal measures			
Gestational age/weeks	28.7 (26–33)	39.4 (37–42)	$p<0.001$
Birth weight/g	997.4 (700–1240)	3394.8 (2212–4648)	$p<0.001$
9th year measures			
Adjusted age*/yr	8.6	8.7	
Height/cm	127.8	132.1	$p<0.05$
Full-scale IQ	93.2 (49–126)	116.7 (75–145)	$p<0.0001$

*The adjusted age is measured from the expected due date, not the date of birth.

◆   How many weeks before term were the pre-term babies born (approximately)?

◆   About 11 weeks. The mean gestational age of the pre-term babies was 28.7 weeks which is 11.3 weeks before term, at 40 weeks.

◆   How many weeks premature was the most pre-term baby?

◆   The most premature baby was born at 26 weeks, so it was born 14 weeks before term.

The average IQ score in their 9th year, for those babies who were born pre-term was significantly less than the average IQ score of the similarly aged full-term babies.

◆   Does it follow that a baby born prematurely will have a low IQ compared to a baby born at full term?

◆   It does not follow that a baby born prematurely will have a low IQ compared to a baby born at full term.

Look at the range of values for IQ. Within this relatively small sample of 8(+)-year-olds, some who were premature babies achieved IQ scores well above the scores of some who were born at full term. (Some pre-term babies had IQs of over 120, while some full-term babies had IQs as low as 75.)

In the same study, the authors used MRI scanning to measure the sizes of various parts of the brain. Whilst the basal ganglia, amygdala and hippocampus were significantly smaller in those born pre-term, their ventricles were significantly larger.

This study illustrates an important aspect of development. Factors (prematurity in this case) that impact on development can alter both physical characters (height and brain size) and behavioural characters (IQ) for some considerable time after the impact (8+ years in this case). (Prematurity is revisited in Section 3.4.2.)

Prematurity is a rather general environmental variable involving lots of different factors. The next section considers a very specific environmental variable, light, and its impact on the development of vision and the visual system.

3.3.2   Plasticity and permanency

The visual system relies on, amongst other things, the exquisitely precise connections between the retina, the lateral geniculate nucleus of the thalamus and the visual cortex (see Book 1, Section 3.4.6). In precocial organisms, i.e. those born or hatched able to see and move about, such as horses and ducks, these connections arise in complete darkness. The information necessary to establish the connections must therefore be in the genome. The question then arises as to whether the environment, visual stimulation in this case, could influence the formation of these connections? To answer this question, Blakemore and Cooper (1970) dramatically altered the visual environment, and hence the visual input their subjects received.

◆   Do you think Blakemore and Cooper used a precocial organism, or one that is relatively immature at birth, an altricial organism?

◆   They reasoned that they would be more likely to influence the development of the visual system in an organism whose visual system was still developing. So they used an altricial organism, the cat *Felis catus*.

They discovered that kittens raised for the first three months of life in a visual environment consisting solely of vertical (or near vertical) lines were subsequently unable to see horizontal lines and they bumped into horizontal bars. Similarly, kittens raised for the first three months of life in a visual environment consisting solely of horizontal (or near horizontal) lines were subsequently unable to see vertical lines and they bumped into vertical bars. Neurophysiological examination of the visual cortices of cats reared under such strange visual conditions revealed a corresponding change at the cellular level. It is possible to record from individual neurons in the visual cortex whilst presenting visual stimuli to the animal. If the visual stimulus is a line, then it is found that neurons respond differently to different orientations of the line. In cats reared in a normal visual environment, some 50% of neurons give a noticeable response to a vertical line. However, in cats reared in a vertical visual environment, some 87% of neurons give a noticeable response to a vertical line. Thus, there is an increase in the number of neurons that respond to vertical lines. Similarly, in cats reared in stroboscopic light and therefore unable to see continuous movement, only 10% of neurons in the visual cortex responded to a moving line, compared with 66% of such neurons in cats reared in a normal visual environment.

These differences in visual responsiveness reflect underlying differences in the micro-anatomy of the visual system. Axons in the visual pathways have made slightly different connections with their targets, under the influence of the unusual visual stimulation.

The visual environment therefore does indeed influence the development of the visual system. Furthermore, the cats remained visually impaired for the four years during which they were tested after their visual environment had been returned to normal. The effect is normally considered to be permanent.

Having established that the developing nervous system is responsive to environmental factors, i.e. it is plastic, the next question to consider is whether there is a time limit on this **plasticity**. Does it matter, for example, *when* the visual input is altered for these changes in the visual cortex to appear? In other words, is there a particular period in development which is sensitive to particular environmental stimuli?

3.3.3 Sensitive periods

The steroid hormone testosterone plays a major role in the development of mammals. In particular it is instrumental in causing differences between males and females. One well explored difference concerns play-fighting in young rodents. In the rat, play-fighting is a sequence which begins when one animal pounces on another. The pounce is followed by wrestling and/or boxing and the play-fight usually finishes with one animal on top of the other. A similar sequence of play-fighting is seen in young Rhesus monkeys. Throughout adolescence, the males of both species initiate and become involved in play-fights more frequently than females. The females do play-fight, and the behavioural components of their play-fighting are the same as those used by males, but the female adolescents play-fight less frequently.

To examine the role of testosterone in the development of play-fighting two main procedures are used. The first involves the removal of the principal source of the hormone. In this case the testes are removed from males. The second involves the administration of the hormone of interest, testosterone in this case. The hormone can be administered either to females, who naturally produce relatively little testosterone, or to castrated males, i.e. males whose testes have been removed.

◆   In general terms, what does the first procedure, the removal of the principal source of the hormone, reveal?

◆   The first procedure reveals how development proceeds in otherwise normal males in the absence of the hormone of interest.

The results of this procedure are clear. The play-fighting of male rats castrated soon after birth was reduced in frequency, to a frequency similar to that seen in female rats. This result is highly suggestive of the role of testosterone, but it is not conclusive.

◆   Why is the result from the first procedure not conclusive?

◆   It is not conclusive because there may be other factors produced by the testes that affect play-fighting. Removal of the testes also removes these other factors.

Many developmental studies take as a starting point the effect of the removal of a key stimulus on the development of a particular behaviour or **character** (sometimes called a trait). However, such studies must be treated with caution because of the very real possibility that several stimuli are removed, along with the stimulus in question.

◆   How can the role of testosterone be confirmed?

◆   The role of testosterone can be confirmed by administering testosterone, i.e. by using the second procedure.

Testosterone administered soon after birth to female rats or to castrated males results in a level of play-fighting similar to that of normal males.

In both the procedures described above the phrase 'soon after birth' has been used. The phrase is important because if the procedures are carried out too long after birth, they have no effect on play-fighting. Gestation lasts 21 days in the rat. Testosterone can only exert its effects on play-fighting from about three days before to six days after birth. Once this neonatal period has passed, castration of males or testosterone injection of females has no effect on play-fighting. So the answer to the question posed at the end of the previous section, Section 3.3.2, is clear – there is a particular period in development which is sensitive to particular environmental stimuli. The time restriction or developmental window during which a stimulus can exert an influence on the organism is referred to as a **sensitive period**.

The effect of testosterone, or lack of it, on the developing organism during the sensitive period is permanent.

Testosterone is said to masculinize the organism, because it promotes those behaviours predominantly seen in males. In its absence, the organism is de-masculinized. There are also feminine behaviours, such as the proceptive behaviours seen in the rat, of hopping, darting and ear wiggling prior to sexual activity. They are seen in males castrated neonatally. These behaviours are not seen in the intact male, nor in the female treated with testosterone neonatally. In this case testosterone de-feminizes, by reducing the frequency of proceptive behaviours.

The examples discussed so far show that rather dramatic changes in environmental factors, be they external factors, outside the body, or internal factors, inside the body, exert an effect on the course of development, and have a permanent effect. However, dramatic environmental events during development, with the notable exception of birth, are rare. Much more common are subtle changes in environment. The next section considers whether subtle differences in environmental factors can also have discernible effects on development.

Summary of Section 3.3

The developing organism is nudged onto different developmental paths by the environment in which it finds itself. Thus the experience of being premature, or of experiencing only horizontal visual stimuli, or of experiencing testosterone affects the kind of individual the organism becomes. And the effect of the environmental factors is both profound and enduring; the individual will, quite literally, never be the same again.

3.4   Confounded variables: sensitive skin

A moment's reflection will convince you that parental behaviour differs from one family to another. The effect that different parental styles have on the development of the recipient offspring is very difficult to establish. In part this is to do with the host of other differences between families, not least their genetics and socio-economic status. But also the difficulty arises because to determine cause and effect requires prolonged and intrusive observations of a sort that is not possible on human families. However, such observations are possible in other animals and they shed light on the impact of parenting on the developmental path.

3.4.1   Licking/grooming–arched back nursing

Rat mothers perform a number behaviours towards their pups: they build a nest for their pups, keep them in it and occasionally lick them and nurse them. (Rat fathers have a parental role too but it is not essential and the experimental set-up is simplified by his absence.) Licking occurs predominantly at the time when the dam arches her back and nurses her young, allowing a composite behaviour of licking/grooming–

arched back nursing to be identified and recorded. If licking/grooming–arched back nursing is recorded for a number of rat mothers, consistent differences emerge. Some mothers perform licking/grooming–arched back nursing at a high frequency, others at an intermediate frequency and still others at a low frequency. Furthermore, these differences between mothers are consistent from one litter to the next. So if a dam has a high frequency of licking/grooming–arched back nursing with her first litter, she will also have a high frequency of licking/grooming–arched back nursing with her second litter. And although there are consistent differences in frequency of licking/grooming–arched back nursing between mothers, the overall amount of time in contact with the pups does not differ between mothers.

◆   Why is this last statement important?

◆   It means that the rat pups in the low-frequency licking/grooming–arched back nursing group are not being neglected; they simply receive a different pattern of maternal care.

The difference in maternal behaviour is correlated with a difference in the behaviour of their offspring. A commonly used and easily administered test of rat behaviour is the Open Field Test. A relatively large (2 m diameter), well-lit, white, circular arena is used. The floor is marked with a grid of lines, so movement can be quantified. A rat (the procedure invariably uses solitary animals) introduced to the arena can move about a lot or a little and can hug the sides or venture into the exposed central area. The test lasts either five or ten minutes. Rats differ in how they behave in an Open Field Test.

Rats raised by mothers exhibiting a high level of licking/grooming–arched back nursing (lg–abn), tested in an Open Field, spent significantly more time in the central area, compared to rats raised by mothers exhibiting a low level of licking/grooming–arched back nursing. The question is, does the difference in nursing cause the difference in Open Field behaviour? Correlations are intriguing but they do not reveal anything about causes. It could be, for instance, that in the present example, there is a single underlying cause, passed from mother to offspring, that affects both maternal behaviour and Open Field behaviour. If this is the case, then maternal behaviour may or may not affect Open Field behaviour. To answer the question about causes posed above, a fostering experiment is needed.

Fostering separates offspring from their biological mother, and allows the influence of the foster mother on the development of the fostered pups to be examined. Francis and colleagues, the authors of the experiment described below, were very careful to minimize the disruption to the litters that fostering can sometimes cause (Francis *et al.*, 1999). Only two pups were fostered into or out of any of the litters, which all comprised 12 pups. All dams received two pups and all dams had two pups removed; a fostering variant called cross-fostering.

Figure 3.9 shows how the pups were fostered between the dams.

The hypothesis under test is whether rats which experience a high lg–abn frequency as pups differ in their Open Field behaviour from rats which experience a low lg–abn frequency as pups.

◆   What were the two conditions used in this experiment?

◆   One condition was high lg–abn frequency and the other condition was low lg–abn frequency.

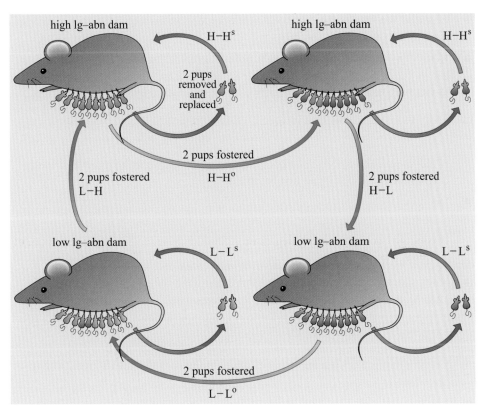

Figure 3.9 Diagramatic representation of cross-fostering. (lg–abn is licking/grooming–arched back nursing.) Two pups were removed from each litter and two pups were added to each litter. (L–L° means the pups were fostered to another low lg–abn dam; H–H° means fostered to another high lg–abn dam; L–H means pups born to a low lg–abn dam were fostered to a high lg–abn dam; H–L means pups born to a high lg–abn dam were fostered to a low lg–abn dam). In addition, two pups were removed and replaced in the same litter. (L–L^s means fostered to the same low lg–abn dam, whereas H–H^s means fostered to the same high lg–abn dam.)

◆ If dam behaviour during infancy is the primary determinant of later Open Field behaviour, how would you expect rats, born of low lg–abn frequency dams, but reared by a high lg–abn dams to behave in the Open Field?

◆ You would expect rats, born of low lg–abn frequency dams, but reared by a high lg–abn dams to behave in exactly the same way as rats, born of high lg–abn frequency dams, and reared by high lg–abn dams. Moreover, they would spend more time in the central area of the Open Field than would rats reared by low lg–abn dams.

The results of this experiment are presented in Figure 3.10.

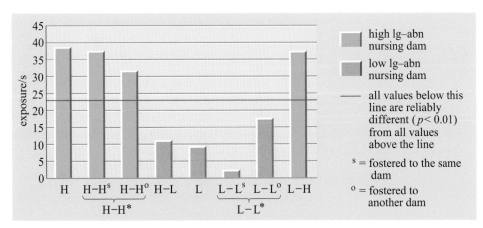

Figure 3.10 Effect of nursing dam on the time (in seconds) spent in the central area of an Open Field (exposure) during a five-minute test. (The * indicates fostered from either the same dam or a different dam of the same lg–abn habit, so L–L* means either L–L° or L–L^s.)

◆ Describe the results for rats born of low lg–abn frequency dams but reared by high lg–abn dams.

◆ Rats born of low lg–abn frequency dams but reared by high lg–abn dams (L–H bar in Figure 3.10) showed no difference in their Open Field behaviour compared to rats born of high lg–abn frequency dams and reared by high lg–abn dams (H–H* bars). However, they spent more time in the central area compared to rats born of low lg–abn frequency dams and reared by low lg–abn dams (L–L* bars).

◆ Is this the result expected if maternal behaviour during infancy is the primary determinant of later Open Field behaviour?

◆ Yes it is.

Having shown that maternal behaviour during infancy is a major determinant of later Open Field behaviour, the same authors went on to demonstrate that it is also a major determinant of later maternal behaviour. When female rats born of low lg–abn frequency dams but reared by high lg–abn dams were themselves mothers, their maternal behaviour was the same as their *nursing* dam. Thus these two characters (frequency of lg–abn; Open Field behaviour) run in families, they are familial, but are environmentally as opposed to genetically determined.

◆ Recall a familial disease mentioned earlier in the course which is genetically determined.

◆ You may have recalled Huntington's disease (see Book 1, Section 1.1.3) or narcolepsy (Section 2.5.3 in this book).

Note that the fact that a disease is familial, does not necessarily mean that it is genetic. We shall return to this subject later in the chapter.

3.4.2 Small babies: the sequel

Evidence that some of the consequences of being born prematurely were enduring was discussed in Section 3.3.1. However, what emerges from Section 3.4.1 is that the quality of maternal care can alter the course of development. So the question arises as to whether the course of development of premature babies can, likewise, be altered by the quality of maternal care? The question is straightforward; obtaining an answer is not.

◆ Using the information from the previous section, identify three things that make answering the question just raised particularly difficult.

◆ One reason is that babies cannot be randomly fostered between mothers with different maternal skills. A second reason is that there are other members of the families to consider (fathers or partners cannot be removed from the situation to make the investigation less complicated). Thirdly, prolonged, intrusive observation is not possible, making evaluation of human maternal behaviour problematic.

The first two difficulties can be overcome, at least in part, by carefully selecting participants, so that the babies and their families match as closely as possible on as many criteria as possible, except for the crucial variable of maternal care. One investigation that achieved this was undertaken by Feldman and colleagues (Feldman *et al.*, 2002). Feldman found two hospitals whose catchment area was similar and level of care was similar. (The details of which criteria were matched are not important to the present discussion, but some are presented in Box 3.1.)

Box 3.1 Criteria for comparison used by Feldman *et al.* (2002)

'Infants in the two groups were matched for gender, birth weight, gestational age, and medical risk. … All families were middle class … and matched for maternal and paternal age, education, parity … and maternal employment. Mothers were all married to the child's father and in all families at least one parent was employed. None of the mothers reported smoking or using drugs during pregnancy.'

'The nurseries in the two hospitals were level 3 referral centres with a comparable number of admissions, case mix, physician level and experience and nurse–patient ratios. … The physical environment in both nurseries, as to light and noise levels, was comparable. In both units, parents had unlimited privileges and were encouraged to participate actively in infant care routines.'

The third problem was overcome, to some extent, by finding two hospitals whose regime differed on the aspect of maternal care of interest: one hospital had a policy of kangaroo care, the other did not. Kangaroo care (so called because of a superficial similarity to a kangaroo carrying her baby in her pouch) is full body, skin to skin contact between mother and infant. The infant would be removed from its incubator, undressed (except for a nappy) and placed between its mother's breasts. Infants remained attached to a cardiorespiratory monitor and were observed by nurses. The criteria for inclusion in the study were that the parents agreed to give their premature infants at least one hour a day of kangaroo care for at least 14 days, before discharge from hospital. These infants were matched with premature infants in the other hospital who received a comparable amount of maternal care, over a similar period of time, but did not receive kangaroo care.

◆　Which was the experimental and which the control condition?

◆　The experimental condition is the one in which the infants received the kangaroo care, whilst the control condition is the one in which the infants did not receive kangaroo care.

To evaluate the effect of kangaroo care on the babies, infant cognitive development was assessed at 6 months of age. This was an adjusted age. It was not six months after they were born, but six months after the date on which they would have been born had they been born at full term.

◆　Why might the use of an adjusted age be important when considering cognitive development?

◆　The adjusted age allows a comparison to be made between babies who have had the same period of time in which to develop, irrespective of the length of gestation.

The cognitive scales used to evaluate infant development have been derived from studies of full-term babies; so the adjusted age allows meaningful comparisons to be made between babies born at different gestational ages.

Infant cognitive development was assessed by a trained psychologist who did not know to which groups the children had been assigned. The psychologist used a standard test (The Bailey Scales of Infant Development) which produced values for both psychomotor development, a psychomotor index (PMI) and for mental development, a mental development index (MDI). The results are presented in Table 3.2.

Table 3.2   Effect of kangaroo care (KC) during infancy on two measures of infant development at 6 months of adjusted age.

	KC ($N = 66$) mean	SD	Control ($N = 67$) mean	SD	Probability the difference between KC and the control mean arose by chance
Birth weight/g	1245	328	1289	358	(no difference)
Gestational age/weeks	30.4	2.5	30.8	3.0	(no difference)
Mental development index (MDI)	96.39	7.23	91.81	9.80	$p < 0.01$
Psychomotor index (PMI)	85.47	18.42	80.53	13.33	$p < 0.05$

Look at Table 3.2 and compare the two conditions. Notice first that at birth, the babies in the two groups were of very similar weight and very similar gestational age.

◆   What effect did kangaroo care have on mental performance at 6 months?

◆   The mental development index is significantly higher for those infants who received the kangaroo care than for those infants who did not; kangaroo care improved mental performance.

◆   What effect did kangaroo care have on psychomotor development at 6 months?

◆   The psychomotor index is significantly higher for those infants who received the kangaroo care than for those infants who did not; kangaroo care improved psychomotor performance.

This example illustrates that one environmental intervention, kangaroo care, can alleviate to some extent, the effects of another environmental perturbation, prematurity.

The developing organism is subject to an enormous number of environmental factors, each of which exerts its influence on the course of development of the individual organism. Some effects are seismic (e.g. the impact of testosterone) others are more subtle (e.g. the effect of maternal care), but they all contribute to the unique phenotype. What has not been considered so far, is *how* an environmental factor can influence development; to understand that it is necessary to consider how the nervous system itself develops.

Summary of Section 3.4

Two important points emerge from this section. The first is the powerful effect of maternal contact on the development and later behaviour of their charges. In the Feldman study the disadvantages of prematurity were essentially overcome by early maternal contact. The second point is that some behavioural traits really do run in families, without being genetic. The Francis study clearly showed that some aspects of Open Field behaviour were the result of nursing care.

3.5   Mechanisms

Development has so far mostly been assessed in terms of the gross performance, the overt behaviour, of the organism. Relationships have been established between certain environmental events and certain behaviours. In the next three sections, Sections 3.5 to 3.7, there is a major shift in emphasis, from considering behaviour to considering the structure of the nervous system. Essentially, this is doing no more than stepping down a level to look at what is going on inside the organism, in particular with its nervous system, during development. By understanding something of how the nervous system develops, it is possible to explain the mechanism(s) by which environmental stimuli affect behavioural development. The final two sections, Sections 3.8 and 3.9, return again to consider behaviour, except that in these two sections the issue is not the extent to which environmental events can affect the development of behaviour, but the extent to which genes can affect behaviour.

3.5.1 Sexually dimorphic nucleus of the preoptic area (SDN-POA)

As well as affecting behaviour (Section 3.3.3) neonatal testosterone also affects the physical characteristics of some areas of the brain. One of these is a small area of the hypothalamus, the medial preoptic area, which, although small, is much larger in males than in females. This size difference is mediated by testosterone.

There is a group, a nucleus, of neurons in the medial preoptic area of both male and female prenatal rats where the neuronal cell bodies are clustered together at an unusually high density. By about the first or second day after birth, the volume of this high-density nucleus is larger in male rats than in female rats, a difference that persists into adulthood. The density of neurons in this nucleus is the same in both males and females; it is its volume that is different. The high-density area is known as the sexually dimorphic nucleus of the preoptic area, SDN-POA. The difference in size comes about through the action of testosterone on the developing brain.

The effects on the volume of the SDN-POA of a single injection of testosterone into neonatal female rats and of castration of neonatal male rats are shown in Figure 3.11. Note the injection consisted of testosterone mixed with oil, not water or saline solution.

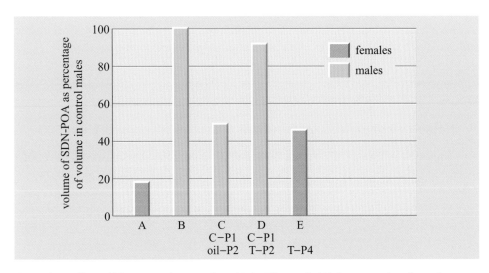

Figure 3.11 Summary of the sex difference in SDN-POA volume and the influence of the administration of postnatal testosterone. SDN-POA volume is expressed as a percentage of the volume of this nucleus in control male rats, 30 days after treatment. Normal female rats are shown in A and normal males in B. In C, males were castrated on postnatal day 1 (C–P1) and given an injection of oil on postnatal day 2 (oil–P2); in D, similar males (C–P1) were also injected with testosterone on day 2 (T–P2). In E, females were injected with testosterone on postnatal day 4 (T–P4).

◆   What effect did neonatal castration (C in Figure 3.11) have on the size of SDN-POA?

◆   The castration of neonatal males reduced the size of the nucleus by about half compared with normal males. (B in Figure 3.11.)

◆   What effect did the neonatal testosterone injection have on the size of the SDN-POA in the females? (E in Figure 3.11.)

◆   The single injection of testosterone into neonatal female rats more than doubled the size of the nucleus compared with normal females, to a size almost half that seen in normal males.

The results from these experiments suggest that the volume of the nucleus is affected by testosterone, but some of the differences still need to be accounted for.

◆   What differences still need to be accounted for?

◆   The volume of the nucleus in castrated males is greater than that of normal females (Figure 3.11 treatment group C compared with group A) and its volume in females treated with testosterone is not as great as that of normal males (treatment group E compared with group B).

◆   In spite of these differences, it could still be claimed that testosterone can account for all of the observed SDN-POA volume differences between males and females. Explain how this is possible.

◆   It would be possible if testosterone exerts some of its effects prenatally. Then these prenatal effects would be missing from the experiments, where all the manipulations were carried out postnatally.

◆   Is there any evidence from these data for a prenatal effect of testosterone on the size of the SDN-POA?

◆   Yes there is. The size of the SDN-POA in males castrated on P1 (that is C in Figure 3.11) is considerably larger than that of the normal females (A in Figure 3.11).

In fact a more prolonged exposure to testosterone, beginning prenatally, does result in female rats with an SDN-POA volume equivalent to that of normal males.

The presence or absence of testosterone during the neonatal period alters the size of the SDN-POA. However, the question of how testosterone exerts its effects remains unanswered. There are essentially two separate problems:

1   How does testosterone alter the volume of neurons in the preoptic area?

2   How does a change in volume of neurons in the preoptic area affect behaviour?

The first question can be answered by considering what testosterone does inside the cell. Testosterone is a steroid hormone, which is important only insofar as steroid hormones can pass straight through cell membranes: no special receptors or transport mechanisms are needed. There are however, special receptors for steroid hormones inside cells. Once inside the neurons of the SDN-POA, testosterone is converted into oestradiol. (*Note*: This may seem perverse, given that oestradiol is one of the main hormones secreted by the mature ovary, and is often characterized as a female hormone. However, the immature ovary, e.g. the ovary of the neonatal female rat, produces very little oestradiol, and what is produced is mopped up by another molecule, alphafetoprotein.)

The oestradiol combines with its receptor, and the combined molecule then interacts with the DNA to produce new proteins. One of the new proteins produced under these circumstances is thought to be what is known as a survival molecule (BDNF) which protects the cell from cell death. Survival molecules are considered further in Section 3.7.

In summary then, testosterone protects the neurons in the SDN-POA from dying.

◆   Does this provide an answer to question 1, posed on the page opposite?

◆   Yes it does. Neurons in the SDN-POA of males are protected and survive, whereas neurons in the SDN-POA of females are not protected and die resulting in a smaller SDN-POA in females.

Question 2 is a little bit more tricky and at present very little can be said about the link between the size of the SDN-POA and behaviour, but two things are clear. The SDN-POA is not part of the sensory pathways to the brain, and neither is it part of the motor pathways from the brain. The differences in size of the SDN-POA must alter, in some way, how information is processed within the brain. The road to resolving this cause–effect relationship still snakes off some way into the distance.

3.5.2   Transcription factors

At various places in this course, reference is made to new proteins being made, or to genes being switched on. (The process of transcription was described in Section 1.5.) The control of gene transcription is a hugely complex area and well beyond the scope of this course. However, it is helpful to know that for any gene to be switched on, for any new protein to be made (or an old protein replaced), an appropriate transcription factor must be present, and it must couple with an appropriate part of the DNA. In biological terms a **transcription factor** is a molecule which attaches to a specific part of the genome and allows a gene to be transcribed. In other words the transcription factor is the signal for the transcription of a particular gene, and the production of a specific protein.

◆   A transcription factor was described, but not named, in the previous section. Of what was it composed?

◆   The transcription factor was composed of the hormone oestradiol and its receptor.

Very often there are several components to a transcription factor, and they must all be present for it to work, for transcription to take place.

3.5.3   Retinoic acid

The retinoic acid story is both distressing and illuminating. It is distressing because with hindsight it is possible to see how the suffering of many people could have been averted. It is illuminating because we now understand much about how retinoic acid works.

Retinoic acid is a natural product of vitamin A. It had been known since the 1930s that a lack of vitamin A, a vitamin A deficiency, led to fetal abnormalities. Subsequent studies in animals showed that an excess of vitamin A also led to fetal abnormalities. Unfortunately that did not prevent a drug that contained retinoic acid coming on to the market and being available to pregnant women. The drug was Accutane® and it was introduced in the 1980s as an effective treatment for intractable acne. Of the children that survived the exposure to this drug in the womb, many were born with defects of the nervous system, in particular of the brain, which was often grossly malformed.

Since the 1980s, the mechanism by which retinoic acid exerts its devastating effect has been uncovered. Two crucial pieces of information have revealed how both a lack and an excess of retinoic acid could be harmful. The first piece of evidence came from studies on the South African clawed toad, *Xenopus laevis*. Having noticed that the natural concentration of retinoic acid was ten times higher at the posterior end of the embryo than at the anterior end, Durston *et al.* (1989) exposed *Xenopus* tadpoles to different concentrations of retinoic acid. Their results are shown in Figure 3.12.

Figure 3.12 (a) Illustration of where retinoic acid (RA) has its effect. (b) Response of *Xenopus* tadpoles exposed to increasing concentrations of RA.

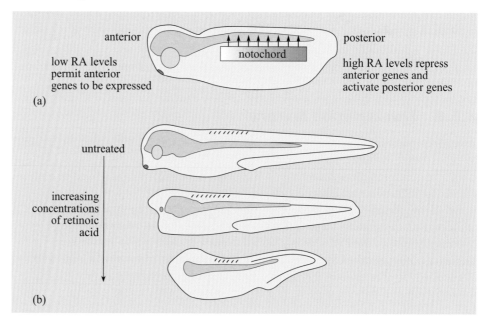

Xenopus embryos exposed to increasing concentrations of retinoic acid fail to develop more anterior body structures, i.e. the brain.

The initial discovery that there is a higher concentration of retinoic acid at the posterior end of the *Xenopus* embryo than at the anterior end led Simeone *et al.* (1991) to surmise that cells responded differently to different concentrations of retinoic acid. To test his idea, he took cultures of human embryonic stem cells and exposed them to either high or low concentrations of retinoic acid.

What Simeone found was that those stem cells exposed to low concentrations of retinoic acid produced proteins normally associated with the anterior part of the embryo. Those stem cells exposed to high concentrations of retinoic acid produced proteins normally associated with posterior parts of the embryo.

◆   What do these results tell you about the action of retinoic acid?

◆   High concentrations of retinoic acid are needed to switch on those genes required for posterior structures, whereas low concentrations of retinoic acid are needed to switch on those genes needed for anterior structures.

Embryos exposed to high levels of retinoic acid because their mothers took Accutane®, had the development of their anterior nervous system inhibited. It turns out that retinoic acid is an intercellular signal, like testosterone, which also passes straight through the cell membrane. Inside the cell, retinoic acid binds to specific receptors (retinoic acid receptors) and the receptor–retinoic acid combination transfers to the nucleus where it acts as a transcription factor (Figure 3.13). That part of the DNA to which the transcription factor attaches is called a response element, in this case the retinoic acid response element (RARE).

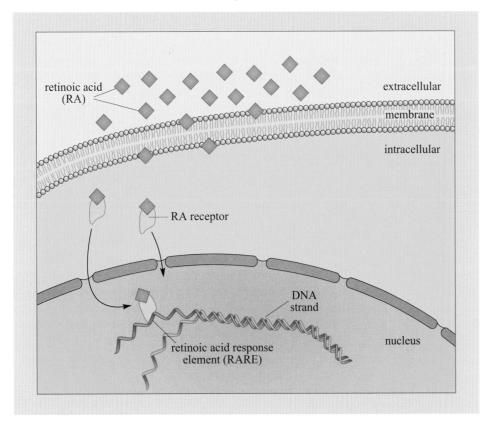

Figure 3.13   Schematic illustration of the signalling and transduction pathway of retinoic acid.

This example illustrates one very important way in which an environmental factor (a drug in this case) can have a direct effect on the early stages of development, by altering the balance of proteins produced in cells. The importance of this balance of cellular proteins is further illustrated in the next section.

Summary of Section 3.5

When oestradiol combines with its receptor inside neurons, the cell produces proteins which protect it from cell death. As a consequence, the male brain, which has oestradiol in its neurons in early life, becomes different from the female brain, which does not have oestradiol in its neurons. Retinoic acid is needed in high concentrations to produce those proteins associated with posterior structures of the embryo, whilst retinoic acid is needed in low concentrations to produce those proteins associated with anterior structures. This is an example of one signal, i.e. one transcription factor, whose effect differs with concentration. Transcription factors are the gatekeepers of the genome, through which environmental factors exert their effects. Profound effects on the physical development of the organism result from interference with transcription factors.

3.6 Axon guidance: the intricacies of neuron growth

Particular nerves, such as those sensory nerves that arise from the nasal retina (the side of the retina adjacent to the nose), cross the midline; other sensory nerves, such as those that arise from the temporal retina do not. This pattern is consistent between individuals, to such an extent that a map of the nervous system is good enough to be used to locate a particular nerve pathway in most people. (See Figure 3.14.)

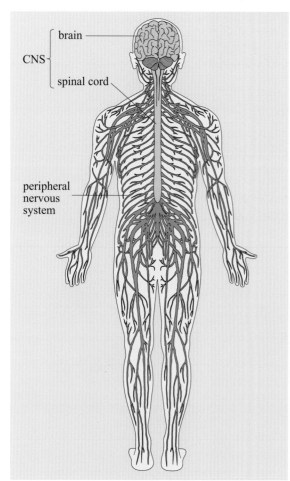

Figure 3.14 The peripheral nerves, as found in most adults.

The precision and consistency of connections raises questions about how these connections come about. The embryo, after all, just has clusters of neuroblasts in various places in the neural tube. As these neuroblasts continue to differentiate into, say, efferent neurons in the motor cortex, they must put out axons that grow and make links to their targets. In the case of a special type of cortical output neuron, Betz cells, the targets are the motor neurons that control the muscles in the arm or leg. Betz cells are either in the left or the right motor cortex, and they innervate motor neurons in the spinal cord on the opposite side of the body. In effect, Betz cells control the muscles in the contralateral limbs. Betz cells do not innervate the muscles at the back of the eye or the motor neurons in the spinal cord on the same side of the body, to control the ipsilateral limbs; they innervate only the motor neurons in the contralateral spinal cord. How do Betz cells do that; how do they cross the midline, from left to right (or right to left) and grow to the contralateral motor neurons and not to, say, the back of the eye? This question of whether to cross or not to cross the midline is addressed later (Section 3.6.3). The next two sections address the question of how do neurons grow (Section 3.6.1) and how do neurons know where to grow (Section 3.6.2)?

There are two main reasons why these two particular questions are addressed here, rather than any of the myriad other possible questions about the development of the nervous system. First, axon growth continues postnatally, and so it is possible to visualize how postnatal environmental events, such as those discussed in earlier sections, might influence axon growth and hence behaviour. Second, axon growth is of major interest to medical science as damage to the nervous system, in particular to the spinal cord, is currently extremely difficult to repair.

3.6.1 The growing axon: growth cones

The growth cone is a small area at the tip of a growing axon (Figure 3.15). As the growth cone moves forward, it adds new material to the cell membrane and so extends the axon. (New axonal membrane is also added at other points along the axon, though to a much lesser extent.)

The growth cone moves forward through the extracellular matrix, as a consequence both of being pushed and of being pulled. The growth cone is pushed by the synthesis of microtubules and the arrival at the growth cone of material transported along the axon from the cell body. The growth cone is pulled by its

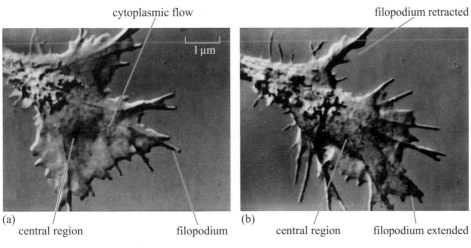

cytoplasmic flow

filopodium retracted

1 μm

(a)

central region

filopodium

(b)

central region

filopodium extended

Figure 3.15   Two video pictures of an *Aplysia* (giant sea slug) growth cone; (b) was taken 7 minutes after (a).

own thin membrane extensions, **filopodia** (singular filopodium). The filopodia extend in various directions from the axon tip and adhere to the surroundings, such as extracellular matrix material, other axons or other cells. Within each filopodium, actin microfilaments are synthesized. These microfilaments are able to contract in a way similar to muscle (actin is one of the components of muscle cells). As the microfilaments contract, so the filopodia contract, pulling the rest of the growth cone along. The rate of advance of the growth cone is about 10–40 μm per hour.

Cytochalasin is a drug that disrupts the formation of actin microfilaments. Cytochalasin also disrupts the formation of filopodia.

◆   What would you expect to be the effect of treating the growth cone with cytochalasin?

◆   Movement of the growth cone would be impaired.

The filopodia are only part of the mechanism by which axons grow, so you would expect disrupting filopodia to slow growth, but not to stop it.

◆   What is the other mechanism by which growth cones move forward, by which axons grow?

◆   The other mechanism is microtubule formation.

Several studies have shown that growing axons treated with cytochalasin meander off course and lose their way.

◆   What does this result suggest about the function of filopodia?

◆   This result suggests that filopodia are important for guiding the growth cone.

One important component in the guidance of growth cones is adhesion; growth cones stick to greater or lesser extents to the surface over which they are growing, their substrate. This adhesion can be illustrated in tissue culture. In tissue culture neurons will extend axon-like processes (called neurites, because in tissue culture axons cannot be distinguished from dendrites) with growth cones at their leading edge. The growth cones adhere strongly to some substrates, e.g. laminin, and to some cells, e.g. guide post cells. The effect of differences in adhesion as a growth cone encounters a guide post cell is shown in Figure 3.16.

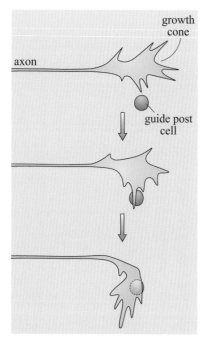

Figure 3.16   Diagram showing how increased adhesion between a filopodium and a guide post cell can direct the growth cone.

It takes just one filopodium to contact the guide post cell for the whole growth cone to change its direction of growth. The growth cone is continuously pulled in various directions by its filopodia. The amount of pull exerted by a particular filopodium will depend to some extent on its size, i.e. how many actin microfilaments it contains, but mostly on its adhesion to the substrate. The filopodium that sticks best to the substrate will pull the growth cone behind it.

◆   What implication does this have for axon navigation?

◆   It implies that the growth cone, and its trailing axon, will follow the path of greater adhesion.

For example, in tissue culture, retinal neurites will grow on two types of extracellular matrix molecules, laminin and fibronectin. (Extracellular matrix molecules are molecules secreted by cells into the surrounding spaces and so are not attached to cells. They also tend not to diffuse very far.) If given the choice, by growing on a surface with alternating fibronectin and laminin stripes, the retinal neurites will grow along the laminin stripes, and not the fibronectin stripes.

3.6.2   Directing the growth cone

The growth of the growth cone has been likened to the progress of a climber. The climber can only go where there are satisfactory hand and foot-holds and where progress is not blocked by physical obstacles (e.g. overhangs or ice). Furthermore, the climber is looking ahead for the best routes, from the current position to the top. Likewise the filopodia on the growth cones are extending outwards, adhering to the best holds and avoiding physical obstacles (e.g. bone or cartilage). This contact-mediated guidance in which immediate (proximal) cues are used is called **chemotactic guidance**. The filopodia are also responsive to more distant cues. The climber uses vision as a distance sense to select the best routes, but this sensory modality is not available to the growth cone. Instead the receptors on the growth cone are responsive to certain molecules that diffuse through the extracellular matrix. These molecules are called **chemotropic factors**, and they may be attractants or repellents. A useful analogy for the interaction between growth cones and attractants, is the sniffer dog. The sniffer dog (growth cone) detects and follows a particular smell (attractant). In so doing, it advances towards the source of a particular smell, the target. For sniffer dogs the target may be explosives or a missing person. For growth cones the target may be a group of muscle cells, a gland or a sensory receptor. The target releases the attractant and the growth cone advances towards the source of the attractant and so towards the target.

3.6.3   Crossing the midline: a case study

Many neurons on the left side of the body make contact with targets on the right, contralateral, side of the body, and vice verse. Crossing the midline is particularly prevalent in descending neurons (e.g. the Betz cells mentioned earlier) and interneurons in the spinal cord which form part of the sensory pathways. (An example of this was seen in Book 1, Section 3.4.6.) However, not all spinal cord interneurons cross the midline: short distance interneurons which influence motor neurons and certain neurons of the spinoreticular tract are ipsilateral. We now look at how the growing axon is guided to cross the midline and become a *commissural axon* or to remain ipsilateral and become a *longitudinal axon*.

In 1993 Seeger reported two *Drosophila* mutants in which axon growth across the midline was affected. In the first, called roundabout or *robo*, axons crossed the midline normally, but then re-crossed and crossed again. *robo* mutants have axons which cross the midline freely. (There is a written convention here: in *Drosophila*, the word in italics, e.g. *robo*, refers to the mutant organism or mutant gene, which does not produce a functioning gene product; the word in normal script, robo, refers to a functioning protein, the gene product.)

The second mutant had axons which never crossed the midline. This mutant is called commissureless or *comm*. *comm* mutants have axons which cannot cross the midline. (See Figures 3.17 and 3.18.)

A fly with both these mutations has axons which cross the midline freely.

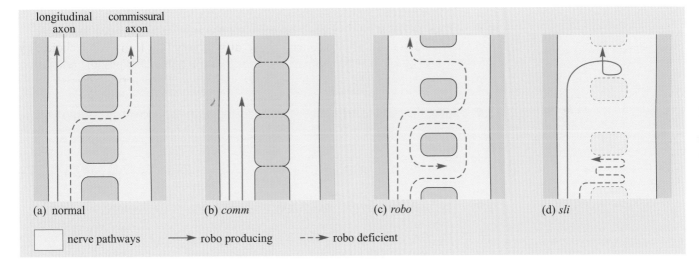

Figure 3.17   Diagramatic representation of longitudinal and commissural axons. (a) In the normal situation longitudinal axons remain ipsilateral, commissural axons cross the midline. (b) In *comm* mutants very few axons cross the midline. (c) In *robo* mutants most axons can cross the midline resulting in enlarged commissural pathways. (d) In *sli* mutants all axons can enter the midline but cannot then leave it, resulting in both pronounced commissures and barely visible longitudinal pathways.

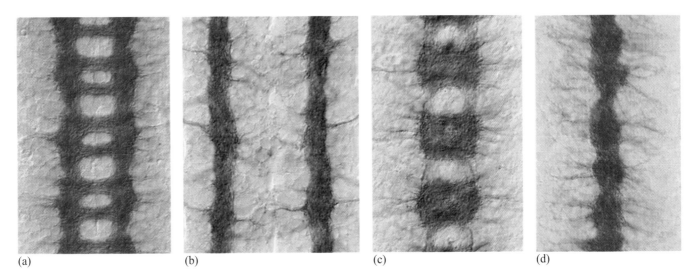

Figure 3.18   The photographs from which the drawings in Figure 3.17 are taken. (a) Normal, wild type, (b) *comm*, (c) *robo*, (d) *sli*.

These data can be explained in numerous ways, which is often the case in developmental studies, but a satisfactory explanation had to wait until 1999 when Kidd discovered an important repellent molecule in the extracellular matrix of the midline (Kidd *et al.*, 1999). The molecule is called sli and it repels most growth cones.

◆ Which mutant growth cones, *robo* or *comm*, does sli not repel?

◆ sli does not repel the growth cones of *robo* mutants since these cross the midline freely.

sli and robo interact. sli is the key, or ligand which interacts with the receptor robo.

◆ What do you think would happen to the growth cone when the molecule sli interacts with the receptor robo?

◆ When sli and robo interact, the growth cone is repelled.

For the growth cone to cross the midline, then, one of two things must happen.

◆ Which two things would allow the growth cone to cross the midline?

◆ Either the midline cells must stop producing sli or the growth cones must stop producing robo.

◆ Given that only some growth cones cross the midline, which do you think is the most likely?

◆ The most likely is that the growth cones stop producing the robo.

If the midline cells stop producing sli then any growth cones could cross the midline. As only some growth cones cross the midline, control must be exerted through the growth cones themselves; thus it is the growth cones that stop producing robo that are able to cross the midline.

Two problems remain. The first is, what is to stop the axons that do cross the midline from crossing it again, as the *robo* mutants do? The second is where does *comm* fit in?

The answer to the first question is that once across the midline, the growth cones of commissural axons begin to produce robo. The mechanism that switches on robo production at the right moment has not been worked out.

The answer to the second question is that comm is an intracellular signal that inhibits (down regulates) the production of robo; in the presence of comm, little or no robo is produced. When comm is not produced (i.e. in *comm* mutants) robo is produced and no growth cones can cross the midline.

There is a parallel medical condition in humans in which cortical axons fail to cross the midline to connect with neurons in the cortex on the opposite side of the brain. The condition, agenesis (pronounced 'a-genesis') of the corpus callosum, is not well understood, but may well result from overproduction of a protein very similar to robo.

Summary of Section 3.6

Growth cones respond to proximal and distal cues. The proximal cues in the extracellular matrix or other cells affect adhesion and result in chemotactic guidance. Distal cues are also in the extracellular matrix but they diffuse through it and result in the growth cone either moving towards the source (attractants) or away from it (repellants). These distal cues are chemotropic cues and can have different effects on different growth cones; what may be an attractant to one growth cone may be repellant to another. It is the interplay of all these factors that results in the axon getting to the correct target.

3.7   Survival and death

There is a huge proliferation of neurons in early life. Even whilst that proliferation continues, some cells, e.g. neuroblasts, stop being able to divide. At some later stage the proliferation itself virtually ceases. It follows that cells switch from being able to divide, to being unable to divide, and that they switch at the appropriate time: the process of cell proliferation is controlled. The details of the control of proliferation are not yet understood and are not considered here. But one feature of the proliferation is worthy of note; it is very wasteful. Over half of the neuroblasts produced die without achieving any functional capacity. This section addresses the question of which neurons live and which neurons die.

3.7.1   Selected to survive: studies of the PNS

Viktor Hamburger carried out a series of classic embryological experiments over a period of about 30 years. He investigated the relationship between the size of target tissue in chick embryos and the size of the pool of neurons that innervated it. His technique was to remove or add target tissue to the tissue which would eventually form a limb, usually the hind limb, and is called the limb bud. A few days later he observed the effect of the tissue addition or removal on the pool of neurons destined to innervate the limb bud. The surgical manipulation (addition or removal of tissue) was carried out on embryonic day 2.5 (E2.5 day) chick embryos and the neural pool examined at E9.5 days (Hollyday and Hamburger, 1976).

◆   Where would he look for changes in the size of the neural pool? (*Hint*: where do the neurons that innervate the limb have their cell bodies?)

◆   There are three possibilities: the ventral horn for motor neurons, the dorsal root ganglion for sensory neurons and the sympathetic ganglion for neurons of the sympathetic nervous system.

Hamburger looked at all of these and his results for the motor and sensory neurons are summarized in Figure 3.19.

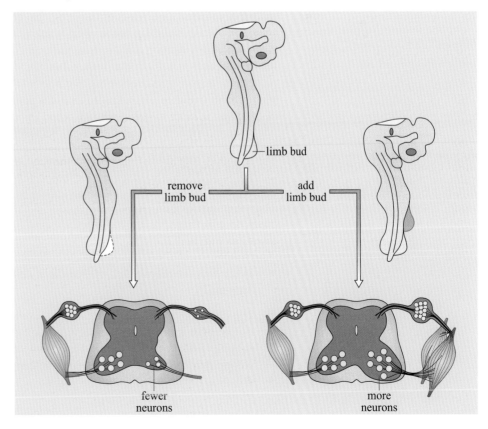

Figure 3.19   Effect of addition or removal of target tissue (a limb bud) on the size of the innervating neural pool (in the ventral horn and dorsal root ganglion).

The results of these and many similar studies led to two hypotheses that have made a significant contribution to the way we think about development. The first hypothesis is that it is the size of the target that ultimately determines the size of the neuron pool that innervates it. In other words, the number of neurons in the neuron pool is reduced or increased to match the size of the target; the target does not increase or decrease in size to match the number of neurons in the neuron pool.

The second hypothesis is about the mechanism by which the matching of neuron pool size to target size is achieved. The quaint hypothesis is that neurons must obtain an elixir from their targets. Neurons that obtain sufficient elixir stay alive. The elixir is secreted by targets, and received by innervating neurons at synapses. This is the neurotrophin hypothesis and it is, by and large, correct.

3.7.2   Selected to die: studies of the CNS

Recent evidence has revealed that during development in mice, cell death occurs in two phases. The first phase is at about E15–E17 days, during neuron proliferation, and will not be considered further here. The second phase at about P0–P5 in the mouse (where P0 is the day of birth) is during the period of innervation.

The evidence for the second phase is well established. Some 50% of retinal ganglion cells die during this phase and here's what happens. Dennis O'Leary (1987) used a long-lasting retrograde tracing dye to look at changes in axonal projections from the retina to the superior colliculus during development of the rat. (This method was also described in Section 2.4.3.) Axons grow from the retina to the superior colliculus, so a retrograde tracing dye injected into the superior colliculus, will be taken up at any of the axon terminals present and transported along the axon to the cell body, located in the retina in this case. When O'Leary injected the dye into the caudal region of one superior colliculus at P12, and looked for dye-filled cells at P14, he found the pattern shown in Figure 3.20a. Most of the dye-filled cell bodies are shown in the nasal region of the contralateral retina. This result is as he expected and reveals the normal way in which the retina and superior colliculus are connected. Figure 3.20b shows the distribution of cells in the retina when the injection was done at P0 (the day of birth) and the retina examined two days later (P2). There are many more filled cells than at P14, and they are scattered over the retina outside the nasal region. This distribution is more widespread than when injection took place at P12. So at P0, when the injection took place, there are retinal ganglion cells scattered throughout the retina that had contacted the caudal superior colliculus, but by P12 these scattered cells have disappeared. Where have all the neurons, shown to be present and filled at P2, gone to by P12? What happens to the missing cells?

The advantage of using a long-lasting dye in these experiments is that the animals could be injected at P0 and the distribution of dye-filled cells examined at P12, as shown in Figure 3.20c.

◆   What does the distribution of dye-filled cells shown in Figure 3.20c suggest has happened?

◆   The distribution looks like that seen in animals injected at P12 and examined at P14 (Figure 3.20a). The small number of filled cells outside the nasal region suggests that most of the retinal ganglion cells that had contacted the inappropriate region of the superior colliculus have gone missing between birth and P12.

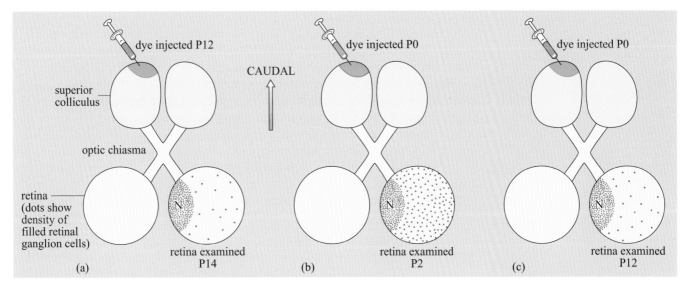

(a) dye injected P12 · superior colliculus · optic chiasma · retina (dots show density of filled retinal ganglion cells) · retina examined P14

CAUDAL

(b) dye injected P0 · retina examined P2

(c) dye injected P0 · retina examined P12

Figure 3.20 Diagrams illustrating the distribution of retrogradely marked retinal ganglion cells in rats of various ages following injection of dye into the caudal region of the superior colliculus at various times earlier in development as indicated. (N is the nasal region of the retina, i.e. adjacent to the nose.)

In fact the missing cells had died. They have not, for example, become concentrated in the nasal region of the retina. Thus the localized projection from the nasal region of the retina to the caudal region of the superior colliculus is due in part to the death of non-nasal retinal neurons which projected to the caudal colliculus.

Some 25–30% of retinal ganglion cells fail to reach their correct targets in the brain and hence die. This leaves the death of a further 25% of retinal ganglion cells to make up the known loss of 50% of retinal ganglion cells. These other neurons innervate the correct targets, so selection among them for those that should die is based on a mechanism for matching the number of neurons to the target size, such as obtaining sufficient elixir as mentioned above.

Our common experience is that healthy organisms die only if they are damaged or injured. Cells certainly do die after damage or injury, e.g. after a stroke, in a process called necrosis. However, cells also use another form of death for which our common experience is not much use. This is programmed cell death, or **apoptosis**, in which the cell makes the proteins that destroy it. The process is called programmed cell death only because each cell has the genes, the cellular mechanism, necessary to cause the cell to die. The signals that start the apoptosis programme are the same kind of signals that set any gene transcription in motion. The difference between these two forms of cell death has been neatly summarized by Sanes 'There is an important difference between a cell that dies gracefully by budding off neat little packages of membrane (apoptosis), compared to one that dies violently by retching catabolic [digestive] enzymes on its neighbours (necrosis).' (Sanes *et al.*, 2000).

3.7.3 Elixirs of the nervous system: neurotrophins

According to Section 3.7.1 axons obtain an elixir from targets at their synapses.

Confirmation that there is indeed an elixir came from a series of events that reveals how much of science really works. Elmer Bucker, working with Hamburger in the mid-1940s, had removed a limb bud from a chick and replaced it with a tumour from the muscle of a mouse – just to see what happened. What happened was that the tumour caused an enlargement of both the sympathetic and sensory ganglia (neuron pools) that normally innervate the limb. Another investigator, Rita Levi-Montalcini picked up on this result and used an extract from the tumour in later experiments on dorsal root ganglia in tissue culture (Levi-Montalcini, 1975). The extract caused a massive increase in the growth of neurites. This led to the extract, or more specifically some component of the extract, being called nerve growth factor. The tumour was treated with various enzymes to find out if different components of the extract were more potent than others. Snake venom was used as a source of some of these enzymes. To everyone's surprise the snake venom itself proved to be a more potent growth-stimulating factor than extracts from the tumour cells. As snake venom is synthesized in the salivary glands, Levi-Montalcini tried an extract from mouse salivary glands. The mouse salivary glands did indeed prove to be a plentiful source of nerve growth factor and a suitable source to use for the difficult process of purification. (Quite why the salivary glands contain nerve growth factor remains an unanswered question.) Rita Levi-Montalcini shared the Nobel prize for Physiology or Medicine for the discovery of nerve growth factor in 1986.

Is nerve growth factor (NGF) the elusive elixir? Three kinds of studies – deprivation studies, enhancement studies, as well as knockout mice studies – proved that it was, for neurons of the sympathetic nervous system at least. Depriving mice of NGF by injecting anti-NGF antibodies (i.e. antibodies that react with NGF) resulted in adult mice almost lacking in sympathetic neurons. Conversely, increasing the amount of NGF in the embryo caused enlargement of sympathetic ganglia. Neurons in these enlarged ganglia were both more numerous and larger. Knockout mice in which the gene encoding NGF has been deleted are lacking in the majority of sympathetic neurons. NGF is indeed the elixir and in time-honoured fashion is called something else, a *trophic factor*. (Also sometimes called a chemotrophic factor, a neurotrophic factor and a neurotrophin.) We will mostly use **neurotrophin** for the rest of this section.

◆　Distinguish between a neurotrophin and chemotropic factor.

◆　A neurotrophin promotes the growth and survival of neurons; a chemotropic factor attracts (or repels) axon growth cones towards (or away from) a target.

The neurotrophin hypothesis was necessary to explain the results of experiments on the limb bud and its innervating motor and sensory neurons. (See Section 3.7.1, where it was first mentioned as a quaint hypothesis.) Nerve growth factor has been shown to affect only the growth and survival of sympathetic neurons, not motor or sensory neurons.

Think about what this last sentence implies.

◆　Bearing in mind the results that the neurotrophin hypothesis was invented to explain, what does the preceding sentence imply?

◆　The sentence implies that there must be at least one other neurotrophin for motor and sensory neurons.

To explain the limb bud transplantation experiments, the neurotrophin (elixir) hypothesis was constructed. The NGF story proves that such elixirs, such neurotrophins, exist. The NGF story also proves that NGF is not the neurotrophin for motor or sensory neurons. There must be other neurotrophins.

The hunt for neurotrophins is intense, but it took ten years for the second (brain-derived neurotrophic factor, BDNF) to be discovered. More recent additions are neurotrophin-3, neurotrophin-4, neurotrophin-6 and neurotrophin-7.

3.7.4 What do neurotrophins do?

Neurotrophins shut down the mechanism of apoptosis. Neurotrophins do this by attaching to receptors in the cell membrane of the innervating axon and activating a cascade of biochemical reactions. The sequence of intracellular events need not concern you. The end result is that enzymes that destroy the cell are not produced and the cell remains alive.

3.7.5 Synaptogenesis

The formation of synaptic connections is an essential property of nervous system development. Synapses are formed between neurons and also with targets that are not part of the nervous system, e.g. muscle. Axon terminals, under the direction of a variety of extracellular cues, grow towards particular targets. Once they arrive at the target, they stop growing and the growth cone changes to form a synapse. As with axon growth, the formation of the synapse is dependent on an interaction between the target and the growth cone. But unlike axon growth, the exchange of chemical signals also operates from neuron to target; the presynaptic terminal releases neurotransmitter and other signals which influence the target. Once formed, the synapse continues in a state of continual change, able to alter its size, its precise location on the target and the extent to which it communicates with the target. These processes of change are essential to allow for adjustment as the animal grows and to allow the accretion of experience, usually known as learning. (This will be discussed in Book 5, Chapter 1.)

3.7.6 Neurogenesis

Brains contain within them the seeds of their own salvation and the seeds of their own destruction. In its early stages, the brain produces vast numbers of neuroblasts as stem cells divide at a huge rate, churning out millions of potential neurons. By birth in humans, this process of neuronal proliferation has virtually stopped. There are, however, some localized areas of the brain, in the olfactory lobe and the hippocampus, for example, where neuronal stem cells survive well into adulthood. These stem cells can produce new neurons. What is not known is what controls the process. Under what conditions, in response to what signals will stem cells become active and start producing new neurons? Under what conditions, in response to what signals will stem cells become inactive and stop producing new neurons? All neurons though also contain the mechanism for apoptosis, for cell death. Under what conditions, in response to what signals will neurons switch on the cell death machinery that will lead to their own destruction? We don't know. But the answers to those questions would open up whole new avenues in the treatment of a variety of brain disorders.

Summary of Section 3.7

This section has sought to illustrate the formation of connections between neurons and their targets by exploring a few examples. The picture that emerges is one of cells at different stages of development subjected to a vast array of signals. These signals are the medium through which environmental factors exert their effects. To some of these signals, some cells respond; to other signals, other cells respond. What a cell, a neuroblast, a growth cone actually does is dependent on the combination of signals around and receptors it produces. Figure 3.21 summarizes the main types of influences on the growth cone of the axon.

The end result is a marvellously choreographed, precision production.

Figure 3.21 Schematic diagram summarizing the main types of influences on the axon growth cone.

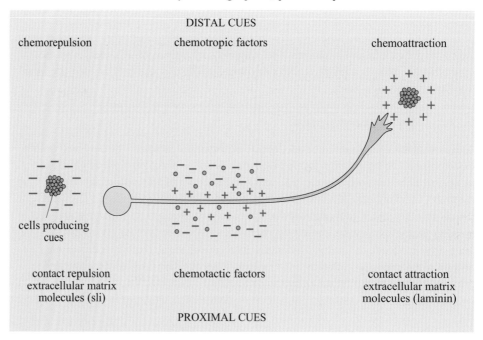

3.8 Genes become you

In the preceding sections many different proteins have been mentioned. These proteins are the receptors, signals, channels, enzymes, transporters, structural components and transcription factors that enable the nervous system to grow and function. Other proteins (e.g. the actin in muscles) are involved in making limbs move and sense organs function. Any and all behaviour is dependent on these proteins. And as each of these myriad proteins is the product of a gene, it follows that any and all behaviour is dependent on many genes. The word used to describe this dependence on many genes is **polygenic**. No behaviour is the product of a single gene. Yet there remains pressure in the popular press and to some extent in society at large, to make just this link; to identify the gene for ***** (you could insert any topical/aberrant behaviour here, e.g. obesity, schizophrenia, etc.). There are two main reasons for this pressure. The first reason is the extraordinary success the medical profession has had in associating particular gene variations (alleles) with certain medical conditions. (Some specific conditions are discussed later in this section.) So, the argument goes, if some conditions, then all conditions; if all conditions, then all behaviours. The second reason for the pressure is that to associate an allele with a condition somehow concludes the quest for the cause of the condition; a bit like a guilty verdict in a murder inquiry. Yet associating an allele with a condition does nothing of the sort, it does not conclude the quest for the cause of the condition. Explanation and treatment require an understanding of what the allele, or more specifically its product, does. Finding the allele is better likened to finding a body, than a guilty verdict; you no longer have a missing person, but you have a long way to go to solve the mystery. The list of the kinds of things that proteins can do is long. In addition, it is also often the case that the product of a single allele, a single protein, can affect many biochemical pathways, and affect many components within the body, a situation known as **pleiotropy** (for example, see Wilson's disease, Section 3.8.1). Finding an allele associated with a particular disease is an important clue as to the causes of the disease, but it does not necessarily solve the mystery.

All of the myriad proteins mentioned in previous sections, the receptors, signals, channels, enzymes, transporters and transcription factors, are products of their respective alleles. New alleles can arise by mutation. When this happens, the new alleles often produce defective proteins by which is meant simply that the protein cannot do its job and as a consequence the phenotype is affected. Amadeus Mozart, Nelson Mandela and Paula Radcliffe may all have had their phenotype affected by defective proteins. But usually the exceptionally gifted are not subject to genetic studies. In contrast, and in the examples that follow, it is the adversely affected phenotype that is subject to genetic study.

Sometimes the absence (occasionally the presence, but more usually the absence) of a particular protein has an effect on an organism that is very conspicuous. Most frequently the effect of the absence is lethal. For those absences that are not lethal, the effect may be conspicuous because it alters the organism's composition, its shape, its response to stimuli, its colour or its behaviour. In such cases the effect of the single missing protein penetrates through the effects of all the other proteins in the body and manifests itself as a conspicuous character or set of characters. The character(s) can be said to be strongly associated with a particular protein and hence a particular gene. And depending on the extent of the effect, the organism may be considered to be simply unusual, or unusual and adversely affected, i.e. diseased. (Unusual domestic animals may go on to found new breeds, or varieties, or strains: consider the range of varieties of dogs, for example.) Where absences have been induced in some way, the process of looking for the effects is called a genetic screen. People who are unusual and adversely affected by being so will turn to the medical profession for help, and it is here that the quest for a gene strongly associated with the character might begin, for example with a family history or twin studies (see Box 3.2).

Box 3.2 Finding genetic correlates in people

There are four main ways in which scientists look for genetic correlates of human characteristics. Those characteristics can be anything from an extra digit on the hands through to behavioural choices (e.g. thrill seekers). These four ways are families, twins, adoption and genetic markers.

Families   Characters that are consistently present in some families and not others, i.e. characters that appear to be inherited, are strong candidates for a genetic search. Tracing family histories and identifying individuals with and without the character can reveal the pattern of inheritance. The genetic basis of Huntington's disease was identified using extensive family pedigrees (Book 1, Section 1.1.3).

Twins   Identical twins have an identical genome. Any character that one twin has, the other twin must also have if the character is strongly associated with a particular gene. There should be what is referred to as a high level of concordance between the twins. Where only some pairs of identical twins share a character, there is said to be discordance and the character is said to be strongly influenced by environmental factors.

Non-identical twins do not share a genome, indeed are no more genetically similar than ordinary siblings, but do share a family environment. Comparisons of concordance between identical and non-identical twins for a character give some indication of the extent to which the character is associated with particular genes.

Adoption   Identical twins share a family environment and so concordance for a character could be due to their shared circumstances rather than their shared genome. Examining concordance where twins have been separated very early in life, usually by adoption into different families, allows the contribution of genome and environment to be distinguished. Note though that adoptive families must meet certain criteria set by the adoption agencies, and so adoptive families may share a number of qualities.

Genetic markers   These are short sections of 'junk' DNA which can be identified biochemically. If two people have one of these short sections in common, then, because of the way in which DNA is inherited, it is likely that they have alleles adjacent to the short section in common too. The short section 'marks' alleles for further investigation.

The three examples that follow in Sections 3.8.1 to 3.8.3 have been chosen to reveal different levels of the complex relationship between genome and phenotype.

3.8.1 Wilson's disease

The effects of a protein that is absent, or present but not doing its job, may not be evident for many years. This is called **late onset**, and is exemplified by Wilson's disease. Many molecules within the body require small amounts of minerals such as iron, magnesium or copper to function properly. There are mechanisms for absorbing these minerals from the diet. However, in excess, these same minerals can be toxic, as is the case with copper. So there are also mechanisms for getting rid of, excreting, these minerals. In Wilson's disease a protein (unhelpfully referred to as ATP7B), crucial for removing copper from the blood prior to excreting it, is absent, and copper accumulates. Provided one copy of the normal allele for ATP7B is present in the genome, copper excretion can occur. However, if both copies of the gene are mutant alleles, copper excretion cannot occur and Wilson's disease results. Wilson's disease is only evident when there are *two* defective alleles, and is an example therefore of a **recessive** trait. (A **dominant** trait is where only one defective allele is sufficient for a disease to be manifest, e.g. Huntington's disease.)

Copper primarily accumulates in the liver, so the initial symptoms are often to do with liver function. However, copper also accumulates in the CNS and may result in difficulty in writing or speech, trembling, an unsteady walk or loss of mental functions. Symptoms may appear at any time but usually appear during the teenage years. Note that late onset refers specifically to the conspicuous effects, the symptoms, of the disease; the key protein is missing throughout life.

◆ If the protein is missing throughout life, why don't the symptoms appear earlier than the mid-teens?

◆ The missing protein does not directly cause the symptoms, it is the accumulation of copper that directly causes the symptoms. Copper accumulates very slowly, so it takes many years for sufficient copper to accumulate to produce the symptoms.

◆ In what way does Wilson's disease illustrate pleiotropy?

◆ Pleiotropy is where a single gene has multiple effects. The single gene which is damaged in Wilson's disease has effects that are manifest in problems with movement and liver function.

Provided Wilson's disease is diagnosed before any serious damage is done to the brain or, to a lesser extent, the liver, the outlook for the patient, the prognosis, is very good. Drugs (e.g. zinc acetate or D-penicillamine) can be taken to remove excess copper from the body; and once the excess copper is removed, so is the problem. This illustrates very nicely an often misunderstood aspect of genetic diseases. To have a particular allele in the cells of your body, or to put that another way, to carry an abnormal gene, does not mean that the disease associated with that allele is inevitable: environmental intervention, drugs in this case, can control the expression of this genetic disease.

3.8.2 Lissencephaly

Lissencephaly, literally meaning 'smooth brain', is characterized by the absence of sulci and gyri, and by a four-layered cortex, instead of the usual six layers, with the majority of cortical neurons in layer four (Figure 3.22). Babies born with lissencephaly have a very poor prognosis; the disease proving lethal before their second birthday. Behaviourally, lissencephaly results in epilepsy, mental retardation, motor impairment and a general lack of developmental progress.

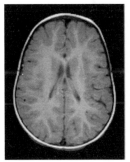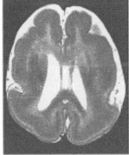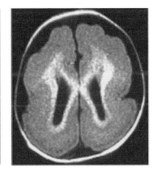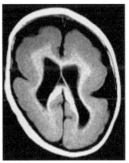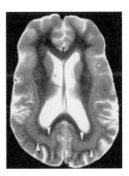

normal

Figure 3.22 Magnetic resonance images of normal and lissencephalic brains. The four lissencephalic brains show a thickened cortex and fewer gyri than is normal.

Lissencephaly results from a mutation in the LIS1 gene[1]. A mutant form of this gene produces a non-functioning protein, which is lethal very early in prenatal development when it occurs on both chromosomes, i.e. when both alleles are defective. In this case the genome is said to be homozygous for it. However, in the heterozygous condition, where both forms are present, a certain amount of development occurs and it has been possible to establish the precise role of LIS1 using heterozygous knockout mice. Heterozygous mice show cortical disorganization consistent with disordered neuronal migration.

During the development of the brain, neurons are born at the ventricular surface, where brain tissue meets the cerebrospinal fluid of the ventricles. Newly born neurons (neuroblasts) move away from the banks of the ventricles towards the outer surface (the pial surface) of the brain. In terms of the number of items (neuroblasts) moving, this is an unimaginably vast migration. The movement of the neuroblasts involves a number of activities, one of which ensures that the nucleus stays within its cell, and doesn't get left behind. This is nucleokinesis. The protein product of L1S1, LIS1P, pulls the nucleus along and so is important in nucleokinesis. In the homozygous mutant condition, the protein is absent/defective and there is nothing to pull the nucleus along, so nucleokinesis and hence cell migration, cannot occur, resulting in the early death of the organism. In the heterozygous condition, some LIS1P is made and so some nucleokinesis can occur, though the exact extent depends on the amount of LIS1P produced and which other genes are present.

Lissencephaly is a **congenital** disease, meaning simply that the symptoms of the disease are present at birth. In contrast to Wilson's disease, lissencephaly is untreatable. This is because intervention would need to happen more or less at conception, be continuous throughout development and target neuroblasts at the ventricular surface.

[1] Note: the nomenclature used here is correct for human genes, e.g. LIS1, and their products, e.g. LIS1P, and differs from the nomenclature used for *Drosophila* in Section 3.6.3.

Although relatively rare, more than 25 different syndromes with abnormal neuronal migrations have been described in humans. Some authorities suggest they account for up to one-third of all cases of severe epilepsy.

3.8.3   Fragile X syndrome

Fragile X syndrome is the final example of a genetic disease considered here.

◆ What does the term 'genetic disease' mean?

◆ Genetic disease means that the symptoms of the disease are associated with a particular allele. Note that the gene may be present, but it would be a faulty version, a faulty allele.

The term genetic disease also means that a protein is not present or is not doing its job properly. It is this alternative meaning that is important because it opens up the question of what the missing/faulty protein is failing to do; of what the function of the normal protein is.

The gene at the centre of fragile X syndrome is FMR1 (an abbreviation for fragile X mental retardation gene 1). The protein at the centre of fragile X syndrome is FMRP (fragile X mental retardation protein).

The symptoms of fragile X syndrome are numerous and varied. Some of the symptoms are physical, and include an elongated face, prominent large ears and a high arched palate. Others are developmental, meaning that certain motor patterns appear later than in unaffected individuals. These developmental symptoms include speech delay, fine and gross motor delay and coordination difficulties. There are behavioural symptoms too, including shyness, sensory defensiveness and mood instability. In addition, learning disability is a common symptom.

The large range of symptoms is due, in part, to two peculiar features of the disease which illustrate the intricacy of the relationship between genes and development. One feature has to do with the nature of the mutation and the complexity of DNA. The fragile X mutation is an *addition*; some DNA is added to the beginning of the FMR1 gene to create a new allele, and if enough DNA is added to the beginning of the gene, the extra DNA prevents the relevant transcription factor from starting transcription. The extra DNA prevents the allele from being transcribed. (The technical term for the addition of the DNA, which you need not remember, is a triplet repeat expansion.) While the embryo is a ball of dividing cells, in 10% of individuals with the mutation, some cells carry an FMR1 allele which can still be transcribed. The upshot is that individuals with fragile X syndrome can have a mosaic of two types of cells, those transcribing and those not transcribing FMR1. As a result, the symptoms can differ between individuals depending on which cells can transcribe FRM1.

The second peculiar feature has to do with the function of the protein product of FMR1, FMRP. The end product of transcription is messenger RNA (mRNA). mRNA leaves the nucleus before being translated into protein. (See Section 1.5.) To leave the nucleus, mRNA must pass through the nuclear membrane, and as you know, membranes are designed to prevent things passing through them. Just as ions need the assistance of protein channels to pass through the cell membrane, so mRNA needs the assistance of protein transporters to pass through the nuclear membrane. To keep cell biologists in work, nature has arranged things so that there are numerous mRNA transporter proteins, each transporter protein assisting only a select group of mRNA molecules out of the nucleus and transporting them for translation. FMRP is involved with regulating the transport and translation of some mRNAs. What this paragraph means is that those select mRNAs which happen to find themselves in cells without FMRP, cannot leave the nucleus.

◆   What is the immediate consequence of mRNA failing to leave the nucleus?

◆   The protein for which the mRNA codes cannot be made.

What these three examples illustrate is that genetic diseases vary enormously in their consequences, i.e. whether they are congenital, lethal, treatable, pleiotropic or what the symptoms might be. Labelling a disease as genetic rules out other potential causes of the disease, such as viruses, bacteria, pathogens and poisons. The medical profession rightly seeks to distinguish between these causes, to allow attention to be focused on the appropriate management and treatment of the disease.

The final section in this chapter considers an example where the quest for a genetic correlate is applied to a psychological problem.

Summary of Section 3.8

Genes do influence development. However, genes do not always determine the developmental path. The prognosis for Wilson's disease is very good, because environmental intervention is possible. The prognosis for lissencephaly will remain poor for the foreseeable future. For many other characters the relationship with the genome is very complex. Searching for genetic correlates to disease will continue to be a major enterprise, but finding such a correlate should be likened to finding an accomplice, not finding a cause.

3.9   Genes and their influence on behaviour revisited

The examples in the previous section followed the traditional medical approach, namely that there is a disease, it can be diagnosed (identified), and the cause of the disease, be it viruses, bacteria, pathogens, genes or poisons, can be sought. This section moves away from the medical arena and into the psychological arena, where the symptoms are behavioural. In this case, the symptoms are socially unacceptable behaviour and to the list of causes just mentioned is added family circumstances and the individual's social situation.

3.9.1   Antisocial behaviour disease

The psychological arena is hugely complex because there are additional issues of responsibility and treatment. Briefly, society takes a more lenient attitude towards the behaviour of someone who is ill (diseased) compared to someone who is well. The diseased person is not fully responsible for their actions ('They can't help it'). Therefore any individual with antisocial (aggressive) behaviour who is diagnosed as having a disease is largely absolved of blame. Having a disease, means, at least in principle, that the disease can be treated. So rather than being regarded as a criminal and punished for antisocial behaviour, the individual with a disease is regarded as a patient and treated for their symptoms. Finding a genetic association with the antisocial behaviour of an individual then does two things. First, it confers a disease status on the individual; the individual has a genetic disease. Second, as the genes are part of the individual's biology, treatment should be focused on the individual's biology, rather than social circumstances, for instance.

In recent years the idea that family circumstances or upbringing (e.g. sexual abuse as a child) can also result in a diseased state in adulthood has been gaining ground.

◆ Is the idea that family circumstances or upbringing (e.g. sexual abuse as a child) can result in a diseased state in adulthood consistent or inconsistent with what you have read so far in this chapter?

◆ The idea that family circumstances or upbringing can result in a diseased state in adulthood is consistent with the material in Sections 3.4.1 and 3.4.2. Environmental factors early in life can have enduring effects on the individual.

The final example in this chapter looks at two very different causes of antisocial behaviour.

3.9.2 Monoamine oxidase A, maltreatment during childhood and later violence

One Dutch family was found to have a history of antisocial (aggressive) behaviour. Genetic studies were conducted and a potential culprit gene MAOA, monoamine oxidase A, identified. The aggressive individuals in the family appeared to have a mutant gene which produced no MAOAP, an enzyme involved in the breakdown of certain neurotransmitters, including serotonin. A knockout mouse model, in which the MAOA gene was inactive, was also found to be aggressive, apparently confirming the role of MAOAP in aggression. However, surveys have shown that not all people showing antisocial behaviour have the mutant gene or abnormal levels of MAOAP.

In parallel with these genetic studies of human aggression, the role of childhood maltreatment as a risk factor in later antisocial behaviour has been studied. Boys who experience abuse, e.g. punitive parenting, are at risk of developing into violent offenders. The abuse increases the risk of later criminality by about 50%. However, most maltreated children do not become violent adults. Just as with the MAOA study above, there is no clear cause and effect relationship.

Terrie Moffitt and Avshalom Caspi and colleagues at the Institute of Psychiatry in London wondered if these two factors, a mutant MAOA allele and maltreatment during childhood, might work in tandem in some way. Moffitt and Caspi turned to the Dunedin Multidisciplinary Health and Development Study underway in New Zealand to test their hypothesis. The Dunedin study has tested a cohort of 1037 people, born in 1972, approximately every two years, with the cohort remaining virtually intact (96%), even after 26 years. The study had collected data on early childhood maltreatment and on aggressive behaviour as adolescents and young adults. All Moffitt and Caspi had to do was determine the MAOA alleles carried by the individuals in the cohort (Caspi *et al.*, 2002). (They tested only the 442 individuals who fitted their selection criteria.)

The hypothesis they were testing was that individuals who had both childhood maltreatment *and* mutant MAOA would be more likely to exhibit antisocial behaviour than individuals who had childhood maltreatment but normal MAOA or no childhood maltreatment and mutant MAOA. One result is presented in Figure 3.23.

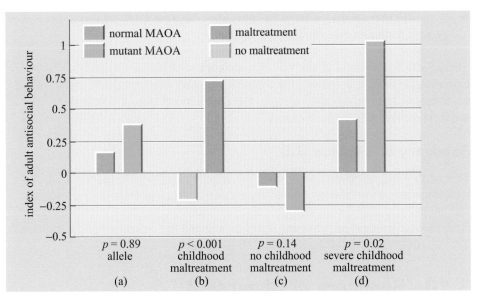

Figure 3.23 Histograms showing the effects of the normal and mutant MAOA alleles and the extent of childhood maltreatment on adult antisocial behaviour. Histograms (a) to (d) all show effects on antisocial behaviour in adulthood. (a) Effect of normal MAOA and mutant MAOA, (b) effect of the absence of maltreatment and severe maltreatment during childhood, (c) combined effect of the absence of childhood maltreatment and either normal MAOA, or mutant MAOA, (d) combined effect of severe childhood maltreatment and either normal MAOA or mutant MAOA. Zero represents the mean level of antisocial behaviour of the people studied: values further away from the mean, above 0, represent higher levels of antisocial behaviour; values further away from the mean, below 0, represent lower levels of antisocial behaviour.

When the sample of individuals is divided into two groups on the basis of their MAOA alleles, and their antisocial behaviour indexes compared, no significant differences emerge (Figure 3.23a).

◆   What does this result mean in terms of the gene of interest and antisocial behaviour?

◆   The result means that the gene of interest does not, of itself, cause antisocial behaviour.

When the sample of individuals is divided into a no childhood maltreatment group and a severe childhood maltreatment group, and their antisocial behaviour indexes compared, a significant difference is found (Figure 3.23b).

◆   What does this result mean in terms of childhood maltreatment and adult antisocial behaviour?

◆   The result means that severe childhood maltreatment often causes adult antisocial behaviour.

The histogram in Figure 3.23c shows that in the absence of childhood maltreatment, antisocial adult behaviour is unlikely to occur, irrespective of whether individuals have normal MAOA or mutant MAOA.

◆   What does the histogram in Figure 3.23d show?

◆   Figure 3.23d shows that those individuals who endure severe childhood maltreatment show significantly less antisocial behaviour as adults if they also have normal MAOA. Conversely, those individuals who endure severe childhood maltreatment show significantly more antisocial behaviour as adults if they also have mutant MAOA.

The data clearly show what the authors refer to as a protective effect against childhood maltreatment of normal MAOA.

◆   What do these data reveal about how the gene MAOA and the environmental factor childhood maltreatment affect development?

◆ These data reveal very clearly the interaction between environmental factors and genetic factors in determining later behaviour.

These data do not reveal in any way that antisocial behaviour is a genetic disease. However, they begin to unravel the complex interplay between genes and environment and perhaps shed light on those who are at risk of developing antisocial behaviour

Summary of Section 3.9

This section has illustrated what has to be done, by way of a long-term study, to yield meaningful information on the relationship between genes and development and the behaviour of the organism. It also illustrates the hugely complex nature of the relationship between genes and development and the behaviour of the organism. Yet this complexity is not the exception, it is the rule.

3.10 Summary of Chapter 3

The chapter began by considering what factors contribute to individual differences. The case was made, with the spiders, and later with genetic diseases, that the genome was very important. Subsequent sections revealed that the external environment (e.g. maternal care, the presence of light) and the internal environment (e.g. hormones and drugs) were very important and that they can both shape and determine the development of the organism. Environmental factors, in the form of hormones and drugs, can play a key role in determining which genes are transcribed. Genes, for their part limit the possibilities. And the proteins that are produced, can influence the organism by influencing cellular processes such as cell migration, axon growth and guidance, synaptogenesis and cell survival.

Learning outcomes for Chapter 3

After reading this chapter you should be able to:

3.1 Recognize definitions and applications of each of the terms printed in **bold** in the text.

3.2 Critically evaluate statements about the influence of the genome on behaviour.

3.3 Explain the ways in which genetic and environmental factors influence the development of the nervous system.

3.4 Provide examples of the influence of genetic and environmental factors on the development of the nervous system and behaviour.

3.5 Discuss possible flaws and short comings in experiments designed to examine the roles of genes and environment in the development of behaviour.

3.6 Explain why different genetic diseases have different outcomes.

3.7 Recognize the significance of experiments similar to those described.

3.8 Explain the role and control of transcription factors.

3.9 Recognize the importance and discuss some aspects of the control of axon growth.

Questions for Chapter 3

Question 3.1   *(Learning outcome 3.2)*

Explain your reasons for agreeing or disagreeing with each of the following sentences.

(a) If a disease has a genetic basis and someone has the (abnormal) alleles for the disease, then that person will have the symptoms of the disease.

(b) If you have the genes for aggression, then you are an aggressive person.

(c) 'His father was a good footballer too. It must be genetic.'

Question 3.2   *(Learning outcome 3.3)*

Use named examples to show how (a) a gene can directly influence nervous system development and (b) an environmental factor can directly influence nervous system development.

Question 3.3   *(Learning outcome 3.4)*

For each of the following statements about the effect of kangaroo care being evaluated after six months, explain why it is true or false.

(a) Six months was the minimum time necessary to demonstrate an effect.

(b) Six months was sufficient time to demonstrate an effect.

(c) If there is an effect after 6 months, then the effect is permanent.

(d) If there is no effect by 6 months, then there is no effect.

Question 3.4   *(Learning outcome 3.5)*

Just as there are different breeds of dog, so there are different strains of rat. To reduce genetic variation, the strains are inbred, meaning simply that offspring are produced only by animals of the same strain. These strains have been bred in captivity for many years and each has particular characteristics. For example, the Fisher 344 strain is highly responsive to novelty and readily secretes large quantities of certain hormones in response to stress (e.g. being restrained). In comparison, Long-Evans rats are considerably less responsive to novelty and secrete far less hormone in response to the same stressor.

(a) Are the differences in characters between the strains familial?

(b) Why might the differences mentioned be thought to have a genetic basis?

(c) What other explanation might there be for the persistence of the differences?

Question 3.5   *(Learning outcome 3.5)*

What technique would you use to find out whether the differences mentioned in Question 3.4 were genetic in origin?

Question 3.6   *(Learning outcome 3.6)*

If a disease has a genetic basis and someone has the (abnormal) alleles for the disease, then that person will have the symptoms of the disease. Give an example where this is the case and an example where this is not the case.

Question 3.7   *(Learning outcome 3.7)*

The following quotation, taken from a leader in *Science* by Constance Holden, is an accurate summary of part of the results presented in a paper (Caspi *et al.*, 2003) in that issue. Read the quotation and then choose, from the list below, the *two* correct statements about the results described.

'Looking back on their records of childhood abuse for the cohort, the researchers found an additional link between 5-HTT gene variants (s-allele and l-allele) and depression: Abuse as a child predicted depression after the age of 18 only in people carrying at least one s-allele. Among the 11% who had experienced severe maltreatment, the double s-allele subjects ran a 63% risk of a major depressive episode. The l-allele participants averaged a 30% risk, regardless of whether they had been abused as children.'

(Holden, 2003)

A   This result demonstrates that abuse as a child causes depression.

B   This result demonstrates that the 5-HTT gene causes depression.

C   This result demonstrates a gene/environment interaction.

D   This result demonstrates that the s-allele of 5-HTT causes depression.

E   This result demonstrates that people with the l-allele will not suffer from depression.

F   This result demonstrates that depression is more likely to occur in people who have the s-allele for 5-HTT and who experienced abuse as a child.

Question 3.8   *(Learning outcome 3.8)*

What is the function of transcription factors and why are they important?

Question 3.9   *(Learning outcome 3.9)*

Comment on the statement that 'all growing axons make straight for their target(s) without hesitation or deviation'.

Question 3.10   *(Learning outcome 3.9)*

Chemotropic and chemotactic factors affect the growth cone. What is the principal difference between chemotropic and chemotactic factors in how they do so?

Question 1.1

(a) Everyone has the same gene for arginine vasopressin.

The sentence is wrong because it implies 'Everyone has the same type of gene (i.e. the same allele) for arginine vasopressin' when people differ in the alleles they have. The sentence can be corrected by removing the word 'same':

'Everyone has the gene for arginine vasopressin.'

(b) The proteome is the full complement of proteins in a cell.

The sentence is wrong because the proteins that have been made and are present in the cell are not the proteome. Proteins that could be made by the cell are the proteome, but only some of them are present in the cell at any one time. The sentence can be corrected by adding the word 'potential':

'The proteome is the full complement of potential proteins in the cell.'

(c) The primary transcription product is mRNA.

The sentence is wrong because the primary transcription product contains the mRNA but also contains unwanted introns. After the introns have been removed from the primary transcription product, what is left is mRNA. The sentence can be corrected by changing the word 'is' to 'contains':

'The primary transcription product contains mRNA.'

Question 1.2

You could have said proteins function as receptors, messengers, structural components, enzymes, contractile elements. (Five functions are given, although only four are required by the question.)

Question 1.3

(a) The template strand is the strand of DNA that codes for the primary transcription product, that contains mRNA.

(b) The exact sequence of bases determines the exact sequence of amino acids in a protein.

(c) The exact sequence of amino acids in a protein determines whether that protein can function or not.

Question 1.4

The words in italics have been replaced to make a more specific statement about translation.

Translation is the process by which *ribosomes* match the sequence of *bases* in *mRNA* to the appropriate *amino acids*, join those *amino acids* together and produce a *protein*.

Question 2.1

A CT scan. An EEG might reveal abnormal brain activity, because of the tumour, but not the location of it. Microiontophoresis would reveal any unusual chemicals in the brain, but only in the place where the cannula was inserted, which might not be anywhere near the suspected tumour. Only a CT scan would reveal the presence and position of the tumour.

Question 2.2

(a) *in situ* hybridization requires only post-mortem brain tissue. Retrograde labelling does require post-mortem brain tissue, but the procedure begins with the living brain.

(b) EEG and MEG are non-invasive; DBS and retrograde labelling are invasive.

Question 2.3

(a) Both autoradiography and immunohistochemistry require samples of tissue.

(b) immunohistochemistry requires antibodies.

(c) autoradiography requires radioisotopes (radioactive material).

Question 2.4

The one major advantage EEG has over all the more recent scanning and imaging techniques is the size of the equipment. It is portable and allows the participant to move around.

Question 2.5

The main reason is that it was considered unethical to require students to take a substance that might be harmful. It may be harmful in itself, but in addition, the motor task may become dangerous if undertaken whilst lacking some elements of motor control. Such an experiment would also alienate those students who for whatever reason do not wish to interact in any way with alcohol.

Question 2.6

The mouse model of narcolepsy enabled a relevant protein, hypocretin, to be identified and its gene manipulated.

Question 3.1

(a) Whilst this may be the case for some genetic disorders (e.g. lissencephaly) it is not the case for all. Wilson's disease and phenylketonurea, for example, can be symptomless provided copper and phenylalanine intake, respectively, are controlled.

(b) The wording of the sentence makes two assumptions: first it assumes that there are genes which affect only aggression, and second it assumes that if someone has those genes, they will necessarily be aggressive. Neither of these assumptions is correct.

(c) If a son has good footballing skills like his father then the most that can be said is that the footballing skills are familial. The skills might have a genetic basis, an environmental basis, or both.

Question 3.2

(a) Examples of alleles affecting the nervous system include *sli*, LIS1, FMR1.

(b) Examples of environmental factors affecting the nervous system include light, testosterone and retinoic acid.

Question 3.3

(a) The statement is false because some effects may have been evident immediately or soon after kangaroo care finished (e.g. heart rate or responsiveness to physical contact), i.e. before the babies were six months old.

(b) The statement is true, as an effect was demonstrated.

(c) The statement is false; from the information provided, there is no way of knowing whether any effects would be evident at any later time. The only way to find out would be to examine the participants at some later time.

(d) The statement is false; noticeable differences could appear after 9 months or after ten years.

Question 3.4

(a) Yes the differences persist through generations and so are familial.

(b) The differences mentioned might be thought to have a genetic basis because the whole point of inbred strains of animals is that they are genetically uniform and genetically different from other strains. Characters that differ between strains are therefore often thought to arise because of the genetic differences between strains.

(c) An alternative explanation for the persistence of the differences might be that differences in maternal behaviour affect the characters in question and affect maternal behaviour. So, for example, any female rat pup reared by an attentive mother rat might become an attentive mother rat in her turn.

Question 3.5

The technique to use would be cross fostering.

Question 3.6

An example where it is the case that someone with the (abnormal) allele(s) for a disease with a genetic basis has the symptoms of the disease is Huntington's disease. Alternatively, you may have mentioned lissencephaly or fragile X syndrome.

An example where it is not the case that someone with the (abnormal) alleles for a disease with a genetic basis has the symptoms of the disease is Wilson's disease. Alternatively, you may have mentioned phenylketonurea.

Question 3.7

The two correct statements are C and F.

The gene in question is for a chemical transporter called 5-HTT (5-HT transporter) that fine-tunes transmission of serotonin, the neurotransmitter affected by the antidepressant Prozac amongst others. The gene comes in two common versions: the long (l) allele and the short (s) allele.

Question 3.8

Transcription factors are essential to gene transcription; without them mRNA cannot be produced nor proteins synthesized. (*Note*: there are many transcription factors, each controlling a specific gene or genes.)

Question 3.9

There is abundant evidence that axons do not grow straight to their targets. They constantly extend and retract branches (filopodia), 'searching' for the appropriate substrate or chemical cues along which to grow. Thus the movement of a growth cone is hesitant. They also deviate around other cells and and they do not grow through tissue that repels them or is too dense.

Question 3.10

Chemotropic factors diffuse away from their source (e.g. the target) through the intervening tissue and affect the growth cone at a distance from their source (i.e. distally). Chemotactic factors are attached to the substrate (e.g. other cells) and affect the growth cones when the growth cones contact the source (i.e. proximally).

Chapter 2

References

Blakemore, R. P. (1975) Magnetotactic bacteria, *Science*, **190**(4212), pp. 377–9.

Lewis, S. (2001) *Trends in Neurosciences*, **24**(10), p. 617.

Luria, A. R. (1986) *The Mind of a Mnemonist*, Harvard University Press.

Sacks, O. (1985) *The Man Who Mistook His Wife for a Hat*, Duckworth.

Siegel, J. L. *et al.* (2001) A brief history of Hypocretin/Orexin and Narcolepsy, *Neuropsychopharmacology*, **25**, S14–S20.

Further reading

Baumeister, R. and Ge, L. (2002) The worm in us – Caenorhabditis elegans as a model of human disease, *Trends in Biotechnology*, **20**(4), pp. 147–8.

Bennet, M. R. (1999) The early history of the synapse: from Plato to Sherrington, *Brain Research Bulletin*, **50**, pp. 95–118.

Carlson, N. R. (2001) *Physiology of Behaviour*, 7th edn, Chapter 5, Allyn and Bacon, Boston.

Freshney, R. I. (1995) Introduction to basic principles, In: *Animal cell cultures, a practical approach*, ed. R. I. Freshney, Oxford University Press, Oxford, pp. 1–14.

Krings, T., Schreckenberger, M., Rohde, R., *et al.* (2001) Metabolic and electrophysiological validation of functional MRI, *Journal of Neurology, Neurosurgery and Psychiatry*, **71**, pp. 762–71.

Nishino, S. and Mignot, E. (1997) Pharmacological aspects of human and canine narcolepsy, *Progress in Neurobiology*, **52**, pp. 27–78.

Chapter 3

References

Blakemore, C. and Cooper, A. (1970) Development of the brain depends on visual environment, *Nature*, **228**, pp. 477–8.

Caspi, A., McClay, J., Moffitt, T. E., Mill, J., Martin, J., Craig, I. W., Taylor, A. and Poulton, R. (2002) Role of genotype in the cycle of violence in maltreated children, *Science*, **297**, pp. 851–4.

Caspi, A., Sugden, K., Moffitt, T. E., Taylor, A., Craig, I. W., Harrington, H., McClay, J., Mill, J., Martin, J., Braithwaite, A. and Poulton, R. (2003) Influence of life stress on depression: moderation by a polymorphism in the 5-HTT gene, *Science*, **301**, pp. 386–9.

Durston, A. J., Timmermans, J. P., Hage, W. J., Hendriks, H. F. J., Devries, N. J., Heideveld, M. and Nieuwkoop, P. D. (1989) Retinoic acid causes an anteroposterior transformation in the developing central nervous system, *Nature*, **340**, pp. 140–4.

Feldman, R., Eidelman, A. I., Sirota, L. and Weller, A. (2002) Comparison of skin-to-skin (kangaroo) and traditional care: parenting outcomes and preterm infant development, *Pediatrics*, **110**, pp. 16–26.

Francis, D. D., Diorio, J., Liu, D. and Meaney, M. J. (1999) Nongenomic transmission across generations in maternal behaviour and stress responses in the rat, *Science*, **286**, pp. 1155–8.

Holden, C. (2003) Getting the short end of the allele, *Science*, **301**(5631), pp. 291–3.

Hollyday, M. and Hamburger, V. (1976) Reduction of the naturally occurring motor neuron loss by enlargement of the periphery, *Journal of Comparative Neurology*, **170**, pp. 311–20.

Kidd, T., Bland, K. S. and Goodman, C. S. (1999) Slit is the midline repellent for the robo receptor in *Drosophila*, *Cell*, **96**, pp. 785–94.

Levi-Montalcini, R. (1975) NGF an uncharted route, in *The Neurosciences. Paths of Discovery*, F. G. Worden, J. P. Swazey and G. Adelman (eds), MIT Press, Cambridge, Massachusetts, pp. 245–65.

O'Leary, D. (1987) Remodelling of early axonal projections through selective elimination of neurons and long axon collaterals, in *Selective Neuronal Death*, G. Bock and M. O'Connor (eds), Ciba Foundation Symposium, Wiley, Chichester, pp. 113–30.

Sanes, D. H., Reh, T. A. and Harris, W. A. (2000) *Development of the Nervous System*, Academic Press, London.

Seeger, M., Tear, G., Ferres-Marco, D. and Goodman, C. S. (1993) Mutations affecting growth cone guidance in *Drosophila*: Genes necessary for guidance toward or away from the midline, *Neuron*, **10**, pp. 409–26.

Simeone, A., Acampora, D., Nigro, V., Faiella, A., Desposito, M., Stornaiuolo, A., Mavilio, F. and Boncinelli, E. (1991) Differential regulation by retinoic acid of homeobox genes of the four HOX loci in human embryonal carcinoma cells, *Mechanisms of Development*, **33**, pp. 215–28.

ACKNOWLEDGEMENTS

Grateful acknowledgement is made to the following sources for permission to reproduce material within this product.

Figures

Figure 2.1 Leigh, E. (Cambridge) 1981 'Collecting microscopes', Christies-Museum of the History of Science, Oxford. Cassell Education Limited; *Figure 2.2 and 2.3* Reprinted from *Brain Research Bulletin*, Vol. 50, No. 2, Bennett, M. R. 'The early history of the synapse', pp. 110 and 112. Copyright 1999, with permission from Elsevier; *Figure 2.4* Mike Stewart/Open University; *Figure 2.5a* Pascal Goetghluck/Science Photo Library; *Figure 2.5b* Wake Forest University Baptists Medical Center website; *Figure 2.6* Reprinted from *Trends in Cognitive Sciences*, Vol. 4, No. 11, Luck, S. J., Woodman, G. F. and Vogela, E. K., 'Event related potential studies of attention', p. 433. Copyright 2000, with permission from Elsevier; *Figure 2.7 Journal of Neurology, Neurosurgery and Psychiatry*, 2001, Vol. 71, pp. 762, 771, Figure 2 with permission from the BMJ Publishing Group. Photograph supplied by Dr Timo Krings, University Hospital of the RWTH Aachen; *Figures 2.8 and 2.9 Journal of Neurology, Neurosurgery and Psychiatry*, 2001, Vol. 71 p. 767, Figures 4 and 5 with permission from the BMJ Publishing Group. Photograph supplied by Dr Timo Krings, University Hospital of the RWTH Aachen; *Figure 2.10* Reprinted from *Mechanisms of Development*, Vol. 92, Sweeney, K. J. *et al.* 'Lissencephaly associated mutations suggest a requirement for the PAFAH1B heterotrimeric complex in brain development', p. 265. Copyright 2000, with permission from Elsevier; *Figure 2.11a and b* Zephyr/Science Photo Library; *Figure 2.12* From *Human Cross Sectional Anatomy* by Harold Ellis. Reprinted by permission of Butterworth-Heinemann; *Figure 2.13a* From J. Nolte, *The Human Brain*, Edition 5, Mosby, courtesy of Dr Nathaniel T. McMullen, Department of Cell Biology and Anatomy, University of Arizona College of Medicine; *Figure 2.13b* Tamamaki Nobuaki; *Figure 2.15a and 2.15b* Vandenberghe Wim *et al.* (2001) 'Subcellular localization of calcium-permeable AMPA receptors...', *European Journal of Neuroscience*, Vol. 14. Blackwell Publishers Limited; *Figure 2.16* Hakan Hall, Karolinska Institutet, Stockholm, Sweden provided the image; *Figure 2.17* (Image provided by Ric Robinson) Robinson, F. R., Houk, J. C. and Gibson, A. R. (1987) Limb-specific connections of the cat magnocellular red nucleus, *Journal of Comparative Neurology*, Vol. 257, pp. 553–77. *Figure 2.18* Emmanuel Mignot, Center for Narcolepsy, Department of Psychiatry and Behavioural Sciences, Stanford University School of Medicine/Sleep Research Center; *Figure 2.19a* From *Flies and Disease*, Vol. 1, 1971, by Bernard Greenberg, published by Princeton University Press, F. Gregor, artist; *Figure 2.19b* Sinclair Stammers/Science Photo Library; *Figure 2.19c* Dawn Sadler/Open University; *Figure 2.20* (top) Hashimoto, K. *et al.* (2001) 'Roles of glutamate receptor S2 subunit and metabotropic glutamate receptor subtype 1 in climbing fiber synapse elimination during postnatal cerebellar development', *Journal of Neuroscience*, December 15, Vol. 21, No. 24. Copyright 2001, Society of Neuroscience;

Glossary terms are in bold. Italics indicate items mainly, or wholly, in a figure or table.